DNA Research Trends

DNA Research Trends

Marius P. Young
Editor

Nova Biomedical Books
New York

For permission to use material from this book please contact us:
Telephone 631-231-7269; Fax 631-231-8175
Web Site: http://www.novapublishers.com

LIBRARY OF CONGRESS CATALOGING-IN-PUBLICATION DATA

DNA research trends / Marius P. Young (editor).
p. ; cm.
Includes bibliographical references and index.
ISBN-13: 978-1-60021-646-6 (hardcover)
ISBN-10: 1-60021-646-3 (hardcover)
1. DNA. I. Young, Marius P.
[DNLM: 1. DNA. 2. Genetic Processes. QU 58.5 D629 2007]
QP624.D17454 2007
572.8'6--dc22 2007010131

Published by Nova Science Publishers, Inc. ✦ New York

CONTENTS

PREFACE

Deoxyribonucleic acid (DNA) is a chemical found primarily in the nucleus of cells. DNA is a long, spiralling molecule that orchestrates the cell's daily operations and provides the genetic blueprint for the physical characteristics of all living organisms. It is the molecule that encodes genetic information in the nucleus of cells. It determines the structure, function and behaviour of the cell. DNA is made up of two complementary strands, the strands intertwine like a spiral staircase to form a structure called a double helix. Subunits, called bases, are the rungs of the staircase. The four nucleotides in DNA contain the bases: adenine (A), guanine (G), cytosine (C), and thymine (T). This volume presents the latest research from around the world.

Chapter I - DNA repair in mycobacteria bears some similarities to that in other bacteria, but there are also some significant differences. The sequencing of the genome of *Mycobacterium tuberculosis* revealed that this pathogenic bacterium possesses recognisable sequence homologues of a number of DNA repair genes involved in pathways that have been well characterised in other bacterial species, including nucleotide excision repair, base excision repair and recombination. However, no genes involved in mismatch repair could be identified and it has subsequently been demonstrated that this pathway is lacking in mycobacteria. Interestingly, an additional pathway of repair not found in *Escherichia coli*, the non-homologous end joining pathway, has more recently been identified in bacteria and the mycobacterial components of this pathway are among those best characterised. A novel mechanism for DNA-damage induced mutagenesis has also been identified initially in mycobacteria utilising a different class of polymerase to those in *E. coli*. Although the classical SOS system of gene regulation in response to DNA damage is conserved and functional in mycobacteria, many of the DNA repair genes whose expression increases following DNA damage are controlled by an alternative system or systems that are yet to be characterised. Interestingly, a number of genes of unknown function are also induced under these conditions, raising the possibility that some of these at least are involved in DNA repair.

Chapter II - DNA endoreplication has been repeatedly reported in protists, plants, arthropods, molluscs, fish and mammals. The same studies indicated that cells possessing endoreplicated genome are generally large-sized and highly metabolically active, suggesting that endoreplication could have a functional significance. Neurons are typically considered as fully differentiated, non-dividing cells containing a diploid genome. Despite this general rule, some papers questioned the validity of this statement and indicated that giant neurons in molluscs, supramedullary and hypothalamic magnocellular neurons in fish and Purkinje cells in vertebrate (predominantly mammalian) cerebellum present DNA contents larger than 2C. Quantitative microfluorimetric evaluation of DNA content in nerve cells of the gastropod molluscs *Planorbarius corneus, Aplysia californica* and *Lymnaea stagnalis* indicated that neuronal DNA contents range between 2C and 200.000C values. This increase in DNA content generally concerns whole-genome duplications, whereas in *P. corneus* endoreplication mainly applies to GC-rich sequences. The second example of endoreplicated neurons is highlighted by the large, clustered neurons located at the boundary between the medulla oblongata and spinal cord in the fish *Lophius piscatorius* and *Diodon holacanthus*. The DNA content of these neurons, evaluated by microfluorimetric methods, range from 4C in small to over 5000C in large neurons. Further experiments with AT and GC specific fluorochromes showed that the increase in DNA content is due to an amplification involving GC-rich DNAs in *L. piscatorius,* whereas a whole-genome endoduplication occurs in *D. holacanthus*. Subsequent quantitative evaluation revealed that also *L. piscatorius* hypothalamic magnocellular neurons, located in the preoptic and tuberal areas, largely exceed 2C. The third example is represented by nuclei of Purkinje cells isolated from the cerebellum. These results have been debated for several years, since contrasting data are present in literature. Up till now, the dilemma remains unsolved, but it is not possible to exclude that a small percentage of Purkinje neurons contain hyperdiploid and tetraploid nuclei, due to additional DNA synthesis. In order to study this topic into more detail, the authors review the available data about endoreplication in invertebrate and vertebrate giant neurons and consider the possible molecular mechanisms responsible for endoreplication. Furthermore, some possible functional significances of neuron endoploidy are discussed.

Chapter III - The National Institute of Standards and Technology (NIST), formerly called the National Bureau of Standards (NBS), has developed over 1300 Standard Reference Materials (SRMs). The Office of Standard Reference Materials defines and describes a SRM as a:

"well-characterized material produced in quantity to improve measurement science. It is certified for specific chemical or physical properties, and is issued by NIST with a certificate that reports the results of the characterization and indicates the intended use of the material. A SRM is prepared and used for the following three purposes: 1.To help develop accurate methods of analysis; 2. To calibrate measurement systems used to facilitate exchange of goods, institute quality control, determine performance characteristics, or measure a property at the state-of-the-art limit; and 3. To assure the long-term adequacy and integrity of measurement quality assurance programs" (http://patapsco.nist.gov/srmcatalog/about/Definitions.htm).

To provide quality assurance in the analysis of human DNA, five Standard Reference Materials - SRM 2390, SRM 2391b, SRM 2392, SRM 2392-I, and SRM 2395 have been developed by NIST and are currently available. Two additional DNA SRMs (# 2394 and # 2399) have been developed and released since the writing of this chapter. Three of these SRMs - the DNA Profiling Standard 2390, the Polymerase Chain Reaction (PCR)-based DNA Profiling Standard 2391b, and the Human Y-Chromosome DNA Profiling Standard

2395 - are intended for use in forensic and paternity identifications, for instructional law enforcement, or non-clinical research purposes. They are not intended for any human/animal clinical diagnostic use. The DNA SRMs, 2392 and 2392-I, are for standardization and quality control when amplifying and sequencing the entire or any segment of human mitochondrial DNA (16,569 base pairs) for forensic identification, disease diagnosis or mutation detection. SRM 2394 is called the Heteroplasmy mtDNA Standard and will permit investigators to determine the limit of their heteroplasmic mutation detection techniques and hopefully, will be useful in promoting the development of new and more sensitive mutation detection methods. Orders and requests for information concerning these SRMs should be directed to: The National Institute of Standards and Technology, Standard Reference Materials Program, 100 Bureau Drive, Mail Stop 2322, Gaithersburg, MD 20899-2322, Telephone: (301) 975-6776, FAX: (301) 948-3730

Chapter IV - Adult stem cells are considerably easier to manipulate than either embryonic stem or embryonic germ cells. At present, the authors believe that of all the mesenchymal stem cells, adipose-derived stem cells (ASCs) are the most promising due to their pluripotency, proliferative efficiency, and low donor morbidity. In this chapter the authors describe experiments using ASCs both *in vitro* and *in vivo*, and in particular, they describe the importance of GFP transgenic mice and rats in these studies. They sought to determine if ASCs harvested from GFP transgenic animals could differentiate into osteogenic, chondrogenic, adipogenic, and neural lineages *in vitro*. In an *in vivo* study, the authors investigated bone and bone marrow regeneration using ASCs harvested from GFP transgenic mice. The results suggest that ASCs harvested from both GFP transgenic mice and rats have the potential to differentiate into osteogenic, chondrogenic, adipogenic, and neural lineages *in vitro*. Additionally, the results indicate that bone and bone marrow can be regenerated subcutaneously, using ASCs harvested from GFP transgenic mice.

Chapter V - Mitochondrial DNA 5178 cytosine/adenine (Mt5178 C/A) polymorphism, which is also known as NADH dehydrogenase subunit-2 237 leucine/methionine (ND2-237 Leu/Met) polymorphism, is reportedly associated with longevity in the Japanese population, and the Mt5178A genotype may possess antiatherogenic properties. Individuals with the Mt5178A genotype may be more resistant to atherogenic diseases, such as myocardial infarction and cerebrovascular diseases, than those with Mt5178C. Moreover, Mt5178 C/A polymorphism has been reported to be associated with blood pressure, intraocular pressure, serum protein fraction levels, and serum lipid levels in healthy Japanese individuals.

In order to investigate the interaction between Mt5178 C/A polymorphism and habitual smoking on serum lipid levels, the authors performed a cross-sectional study in which healthy Japanese male subjects were classified into two groups based on body mass index (BMI): BMI $\leq$ 22 (n = 107); and BMI > 22 (n = 214). Interaction between Mt5178 C/A polymorphism and cigarette smoking was observed with regard to serum triglyceride (TG) levels in BMI $\leq$ 22 subjects (P = 0.026), and with regard to serum total cholesterol (TC) levels and serum low-density lipoprotein cholesterol (LDLC) levels in BMI > 22 subjects (P = 0.019 and P = 0.043, respectively). In BMI > 22 subjects, TC levels were significantly lower in non-smokers with Mt5178A than in smokers with Mt5178A (P = 0.015), while LDLC levels were lower in non-smokers with Mt5178A than in those with Mt5178C or in smokers with Mt5178A (P = 0.047 and P = 0.055, respectively). Although interaction between Mt5178 C/A polymorphism and habitual smoking was not observed with regard to TG levels in BMI > 22 subjects ($P = 0.308$), TG levels were significantly higher in smokers with Mt5178A than in non-smokers with Mt5178A as well as in non-smokers with Mt5178C ($P = 0.011$ and $P = 0.014$, respectively). Taken together, these results suggest that habitual smoking may deprive Japanese Mt5178A (ND2-237Met) genotypic men having BMI of > 22 of their antiatherogenic advantage.

Chapter VI - The major mutagenic and cytotoxic damage to cells caused by ultraviolet light is due to the induction of photoproducts in DNA, principally cyclobutane pyrimidine dimers (CPDs) and (6-4) photoproducts (6-4PP). Most organisms have developed multiple strategies for surviving UV irradiation. These include methods of protection from UV light, pathways for repair of UV-damaged DNA and tolerance mechanisms that allow cells to survive in the presence of unrepaired UV lesions. There is a wide variation in the sensitivities of different organisms to UV, reflecting the different combinations of repair and tolerance mechanisms employed. These mechanisms have been well studied in bacteria and eukaryotes, but are much less well characterised in the third domain of life, the archaea. A useful model archaeon for studies of the response to UV is the sun-tolerant, extremely halophilic, *Halobacterium* sp. NRC-1. This organism is amenable to genetic, bioinformatics and physiological analysis. Furthermore its genome was the first halophilic archaeal genome to be sequenced and well-designed microarrays are available, making it an ideal model archaeon for functional studies and studies of gene expression. A functional analysis of putative repair genes identified from the genome sequence has revealed that bacterial-type nucleotide excision repair, employing *uvrA*, *uvrB* and *uvrC* gene products, is the only dark repair mechanism in *Halobacterium*, in spite of the presence of homologs of eukaryotic excision repair genes such as *rad3* and *rad25*. In natural sunlight photoreactivation is likely to be the most significant repair mechanism, however, and photoreactivation of cyclobutane dimers, but not (6-4) photoproducts, can be accomplished by the Phr2 photolyase. Microarray analysis of UV-induced transcripts has highlighted a response more similar to the eukaryotic than the bacterial response; repair genes are not up-regulated after UV irradiation and there is no evidence of a coordinated SOS response such as that found in bacteria. The most significantly up-regulated genes are *radA1*, the archaeal homolog of bacterial *recA* and eukaryotic *rad51*, and genes encoding subunits of ribonucleotide reductase. These results are discussed in relation to other published work on repair systems in archaea and the response to environmental UV in other microorganisms.

In: DNA Research Trends
Editor: Marius P. Young, pp. 1-45

ISBN: 978-1-60021-646-6

Chapter I

DNA REPAIR IN MYCOBACTERIA

Elaine O. Davis[1] and Lucinda Rand[2]
[1]MRC National Institute for Medical Research, The Ridgeway, Mill Hill, London NW7 1AA, UK;
[2]Dept. Infectious Diseases and Immunity, Imperial College, Hammersmith Campus, DuCane Road, London W12 0NN, UK.

ABSTRACT

DNA repair in mycobacteria bears some similarities to that in other bacteria, but there are also some significant differences. The sequencing of the genome of *Mycobacterium tuberculosis* revealed that this pathogenic bacterium possesses recognisable sequence homologues of a number of DNA repair genes involved in pathways that have been well characterised in other bacterial species, including nucleotide excision repair, base excision repair and recombination. However, no genes involved in mismatch repair could be identified and it has subsequently been demonstrated that this pathway is lacking in mycobacteria. Interestingly, an additional pathway of repair not found in *Escherichia coli*, the non-homologous end joining pathway, has more recently been identified in bacteria and the mycobacterial components of this pathway are among those best characterised. A novel mechanism for DNA-damage induced mutagenesis has also been identified initially in mycobacteria utilising a different class of polymerase to those in *E. coli*. Although the classical SOS system of gene regulation in response to DNA damage is conserved and functional in mycobacteria, many of the DNA repair genes whose expression increases following DNA damage are controlled by an alternative system or systems that are yet to be characterised. Interestingly, a number of genes of unknown function are also induced under these conditions, raising the possibility that some of these at least are involved in DNA repair.

1. Introduction

Damage to DNA can occur as a consequence of internal processes, such as metabolism giving rise to oxidative products, or of external assaults via exposure to genotoxic agents. In either case it is vital to the organism that the damage is repaired in order to avoid mutation or for replication to occur. In many cases the damage affects single bases, for example by alkylation or oxidation. Alternatively, it may take the form of strand breaks, which can be formed directly or arise as a result of processing in the cell. Reflecting the importance of DNA repair for survival, in general organisms possess multiple pathways by which such repair can be mediated. Damaged bases are usually repaired by direct reversal or removed by excision repair, either base excision repair (BER) or nucleotide excision repair (NER). Strand breaks are generally repaired via homologous recombination pathways or in some organisms by non-homologous end joining. For prokaryotes, knowledge of most of these systems is based on studies made in *Escherichia coli*. Here we do not intend to give a detailed review of what is known about each type of DNA repair based on these studies, but rather to provide sufficient information to set the context for discussing what is known about DNA repair in mycobacteria.

2. Direct Reversal

The simplest method of repair is direct reversal of DNA damage. In this mechanism, a specific type of damage to DNA is repaired by a single enzyme composed of a single polypeptide chain, catalysing a one-step reaction to restore the structure of the genome to its normal state (Friedberg *et al.*, 1995). The only known gene in *M. tuberculosis* involved in this process is an alkyltransferase *ogt* (Rv1316c), encoding O^6-alkylguanine-DNA alkyltransferase (Cole *et al.*, 1998). Comparison with other annotated mycobacterial genomes indicates that *ogt* is also present in *M. bovis*, *M. marinum*, *M. avium* subsp. *paratuberculosis* and *M. leprae*. Alkyltransferase activity due to the corresponding gene in *M. bovis BCG* does not alter following exposure to the alkylating agent streptozotocin, whereas in the fast-growing mycobacterial species, *M. smegmatis*, two forms of alkyltransferase activity have been observed, one of which is induced by streptozotocin while the other is not (Wiid *et al.*, 2002). In *M. tuberculosis* the expression of *ogt* is significantly induced in response to H_2O_2 *in vitro* (Schnappinger *et al.*, 2003). This gene is in an operon with *alkA*, a monofunctional DNA glycosylase, and an *M. tuberculosis* strain deleted for both of these genes was hypersensitive to DNA damage *in vitro*, but showed no growth impairment in mice (Durbach *et al.*, 2003). Interestingly, disruption of *ogt* in *M. bovis BCG* resulted in reduced intracellular growth in macrophages (Wiid *et al.*, 2002). Intriguingly, an additional enzyme conferring direct reversal of DNA damage can be identified in the genome sequences of *M. marinum* and *M. avium* subsp. *paratuberculosis* that is not found in *M. tuberculosis, M. bovis* or *M. leprae*: the gene *phr* encoding DNA photolyase. DNA photolyase catalyses the restoral of pyrimidine dimers that commonly arise following exposure to uv irradiation. The observed distribution of this enzyme may be a reflection of the extent of intra-cellular versus extra-cellular existence of these different species of mycobacteria.

Table 1. Conservation of DNA repair genes in fully sequenced mycobacterial genomes

Gene name	M tb	M bovis	M paratb	M marinum	M leprae	Comments
Direct reversal						
Ogt	Rv1316c	Mb1349c	MAP2445	ORF1537	ML1151c	
Phr	x	x	MAP3081	ORF1373	x	photolyase
NER						
UvrA	Rv1638	Mb1664	MAP1341	ORF1603	ML1392	
UvrB	Rv1633	Mb1659	MAP1335	ORF1610	ML1387	
UvrC	Rv1420	Mb1455	MAP1146	ORF1973	ML0562	
UvrD1	Rv0949	Mb0974	MAP0893	ORF1453	ML0153	
UvrD2	Rv3198c	Mb3222c	MAP3297c	ORF785	ML0637	
Mfd	Rv1020	Mb1048	MAP0987	ORF1615	ML0252	
Ercc3	Rv0861c	Mb0884c	MAP0799c	ORF1210	ML2157	
BER						
AlkA	Rv1317c	Mb1351c/ Mb1350c	MAP2443	ORF1536	ML1152c pseudo	
End	Rv0670	Mb0689	MAP4132	ORF113	ML1889c	Nfo
Fpg	Rv2924c	Mb2949c	MAP2994c	ORF1575	ML1658c	MutM
	Rv0944	Mb0969	MAP0889	ORF1444	ML0148 pseudo	possible Fpg
	Rv2464c	Mb2491c	MAP2284c	ORF1043	ML1483c pseudo	Fpg-like
Mpg	Rv1688	Mb1714	MAP1395	ORF1518	ML1351	
Mug	x	x	x	x	x	mismatch specific uracil DNA glycosylase
MutY	Rv3589	Mb3620	MAP0469c	ORF489	ML1920	
Nei	Rv3297	Mb3325	MAP3416	ORF572	x	
Nth	Rv3674c	Mb3698c	MAP0400	ORF348	ML2301c	
TagA	Rv1210	Mb1242	MAP2567c	ORF1819	ML1066	
Ung	Rv2976c	Mb3000c	MAP3016c	ORF1491	ML1675c	
XthA	Rv0427c	Mb0435c	MAP3916c	ORF378	ML1931 pseudo	
MMR						
MutS	x	x	x	x	x	
MutL	x	x	x	x	x	
MutH	x	x	x	x	x	
SbcB	x	x	x	x	x	exonuclease I (XonA)
XseA	Rv1108c	Mb1138c	MAP2686	ORF1816	ML1940	exonuclease VII
XseB	Rv1107c	Mb1137c	MAP2687	ORF1815	ML1941	exonuclease VII
ExoX	x	x	x	x	x	exonuclease X
Recombination						
RecA	Rv2737c	Mb2756c	MAP2848c	ORF1944	ML0987	
RecB	Rv0630c	Mb0646c/ Mb0647c	MAP4092c/ MAP4093c	ORF28	x	
RecC	Rv0631c	Mb0648c	MAP4094c	ORF31	x	
RecD	Rv0629c	Mb0645c	MAP4091c	ORF25	x	
RecF	Rv0003	Mb0003	MAP0003	ORF1741	ML0003	
RecG	Rv2973c	Mb2998c	MAP3009c	ORF1500	ML1671c	
RecJ	x	x	x	x	x	
RecN	Rv1696	Mb1722	MAP1403	ORF1489	ML1360	
RecO	Rv2362c	Mb2383c	MAP2142c	ORF788	ML0633	
RecQ	x	x	x	x	x	
RecR	Rv3715c	Mb3742c	MAP0316c	ORF232	ML2329c	
RecX	Rv2736c	Mb2755c	MAP2847c	ORF1946	ML0988	
RuvA	Rv2593c	Mb2624c	MAP1037	ORF2200	ML0482	

Table 1. (Continued)

Gene name	M tb	M bovis	M paratb	M marinum	M leprae	Comments
RuvB	Rv2592c	Mb2623c	MAP1038	ORF2202	ML0483	
RuvC	Rv2594c	Mb2625c	MAP1036	ORF2198	ML0481	
NHEJ						
Ku	Rv0937c	Mb0962c	MAP0875c	ORF1415	ML2092 pseudo	
LigD	Rv0938	Mb0963	MAP0880	ORF1419	ML2090c pseudo	ATP ligase
Mutagenesis						
PolB	x	x	x	x	x	polymerase II
DinP	Rv3056	Mb3082	MAP3106	x	ML1739 pseudo	Y family polymerase
DinX	Rv1537	Mb1564	MAP1248	ORF1747	ML1197 pseudo	Y family polymerase
UmuD	x	x	x	x	x	
DnaE2	Rv3370c	Mb3405c	MAP3476c	ORF415	ML0416 pseudo	polymerase III subunit
Other potential DNA repair genes						
DinF	Rv2836c	Mb2861c	x	ORF1794	x	
DinG	Rv1329c	Mb1364c	MAP2431	ORF1517	x	probable helicase
HelD	x	x	x	x	x	helicase IV
Lhr	Rv3296	Mb3324	MAP3415	ORF574	ML0716c pseudo	
LigB	Rv3062	Mb3089	MAP3117	ORF1288	ML1747 pseudo	
LigC	Rv3731	Mb3758	MAP0341	ORF176	x	
Nfi	x	x	x	x	x	
RadA	Rv3585	Mb3616	MAP0473c	ORF493	ML0318c pseudo	
SbcC	x	x	x	x	x	SbcCD cleaves hairpins
SbcD	x	x	x	x	x	SbcCD cleaves hairpins
	Rv0269c	Mb0275c	MAP3713c	ORF770	ML2547 pseudo	similar to non-ligase part of Rv0938
	Rv2090	Mb2117	MAP1825	ORF372	ML1335c pseudo	Probable 5'-3' exonuclease
	Rv2191	Mb2214	MAP1930	ORF66	ML0884c pseudo	UvrC domain (Cho-like)
	Rv2719c	Mb2738c	MAP2835c	ORF1977	ML1004	inhibition of cell division
	Rv3201c	Mb3226c	MAP3300c	ORF778	ML0822 pseudo	UvrD domain
	Rv3202c	Mb3227c	MAP3301c	ORF776	ML0821 pseudo	UvrD domain
	Rv3394c	Mb3426c	MAP3487c	ORF401	ML0400 pseudo	similar to DinP
	Rv3395c	Mb3427c	MAP3488c	ORF399	ML0399 pseudo	
	Rv3730c	Mb3757c	MAP0340c	ORF177	ML2342c pseudo	similar to non-ligase part of Rv0938
General						
LigA	Rv3014c	Mb3039c	MAP3054c	ORF1414	ML1705c	NAD ligase
PolA	Rv1629	Mb1655	MAP1322	ORF1621	ML1381c	polymerase I
Ssb	Rv0054	Mb0055	MAP0068	ORF1622	ML2684c	single-stranded DNA binding protein
Regulation						
LexA	Rv2720	Mb2739	MAP2836	ORF1976	ML1003c	

Where a homologue is identifiable in the annotated genomes of *M. tuberculosis* H37Rv, *M. bovis* AF2122/97 (spoligotype 9), *M. marinum* M, *M. avium* subsp. *paratuberculosis* K-10 or *M. leprae* TN the corresponding gene number is given. The abbreviation "pseudo" after a gene number in *M. leprae* indicates that the corresponding locus is a pseudogene that cannot produce functional protein. If no homologue could be identified this is indicated by an x.

3. Nucleotide Excision Repair (NER)

Nucleotide excision repair has been well characterised in *E. coli* and is induced in response to UV damage, which produces cyclobutane pyrimidine dimers and photoproducts, as well as chemical agents that damage DNA either directly by binding to DNA or following activation to DNA reactive species. These include carcinogens, chemotherapeutic drugs (e.g. tamoxifen), and bifunctional chemical reagents that generate interstrand crosslinks (e.g. mitomycin C) damaging both strands of the DNA (Reardon and Sancar, 2005). NER eliminates a broad spectrum of DNA damage and does so by excising an area around the site of the damage (Truglio *et al.*, 2006). Generally in prokaryotes, a 12-13 nucleotide-long oligomer is released that surrounds the damaged area. Excision repair is a three step process: damage recognition, dual incision and release of the excised oligomer, followed by resynthesis and ligation (Friedberg *et al.*, 1995; Reardon and Sancar, 2005). A number of genes have been identified that function in this pathway from studies in *E. coli* and homologues to each of them are present not only in *M. tuberculosis*, but also in *M. bovis*, *M. marinum*, *M. avium* subsp. *paratuberculosis* and *M. leprae* (Table 1), indicating the importance of this pathway of repair.

3.1. UvrABC Excinucleases

Extensive studies of NER in *Escherichia coli* have identified a group of genes termed *uvrABC* (Howard-Flanders *et al.*, 1966; Rupp *et al.*, 1982) and uvrD (Kuemmerle and Masker, 1980; Ogawa, 1970) that function within this repair pathway. UvrAB and C proteins are required for the excision of pyrimidine dimers in *E. coli* and act by endonucleolytic incision of damaged DNA (Friedberg *et al.*, 1995), ultimately resulting in the release of an oligonucleotide. For this reason they are often referred to as excinucleases (Friedberg *et al.*, 1995). In *E. coli*, UvrA dimerises in solution and then associates with UvrB to form the damage recognition complex. Initial identification of potentially damaged DNA is performed by UvrA, but verification of the damage requires UvrB (Truglio *et al.*, 2006), and once bound to the damaged site UvrB unwinds a short segment of DNA and introduces a bend in it, leading to the dissociation of UvrA. UvrC is recruited to the UvrB-DNA complex to catalyse nicking of the DNA backbone, either side of the damaged area (Friedberg *et al.*, 1995; Reardon and Sancar, 2005). Following this, it is thought that the helicase UvrD facilitates the release of UvrC as well as the damaged oligonucleotide from the post-incision complex, allowing DNA polymerase I (the product of the gene *polA*) and DNA ligase (encoded by *ligA*) to complete the repair process (Figure 1).

Homologues of UvrA (Rv1638), UvrB (Rv1633), and UvrC (Rv1420) have been identified in *M. tuberculosis* (Cole *et al.*, 1998), and they would be expected to function in an analogous way in NER. Although, the *uvrA* and *uvrB* genes do not appear to be regulated in a LexA-dependent manner in *M. tuberculosis* they are induced in response to DNA damage (Brooks *et al.*, 2001; Rand *et al.*, 2003). It has been demonstrated that expression of the *uvrA* gene of *M. tuberculosis* is elevated in response to phagocytosis by human macrophages (Graham and Clark-Curtiss, 1999) as well as in granulomas in clinical lungs samples

(Rachman *et al.*, 2006b). An *M. tuberculosis* strain deficient in the *uvrB* gene exhibited severe sensitivity to acidified nitrite and UV light (Darwin and Nathan, 2005). This mutant also showed a somewhat reduced ability to grow in a mouse model of infection as compared to wild-type *M. tuberculosis*, yet was almost fully virulent in mice deficient for gp91^{phox} and iNOS (Darwin and Nathan, 2005). The *uvrC* gene of *M. tuberculosis* is predicted to be required for survival in mouse macrophages (Rangarajan *et al.*, 2002), and is reported to be upregulated in clinical lung samples of TB infected patients (Rachman *et al.*, 2006b). Taken together, these data suggest an important role for NER in *M. tuberculosis in vivo*.

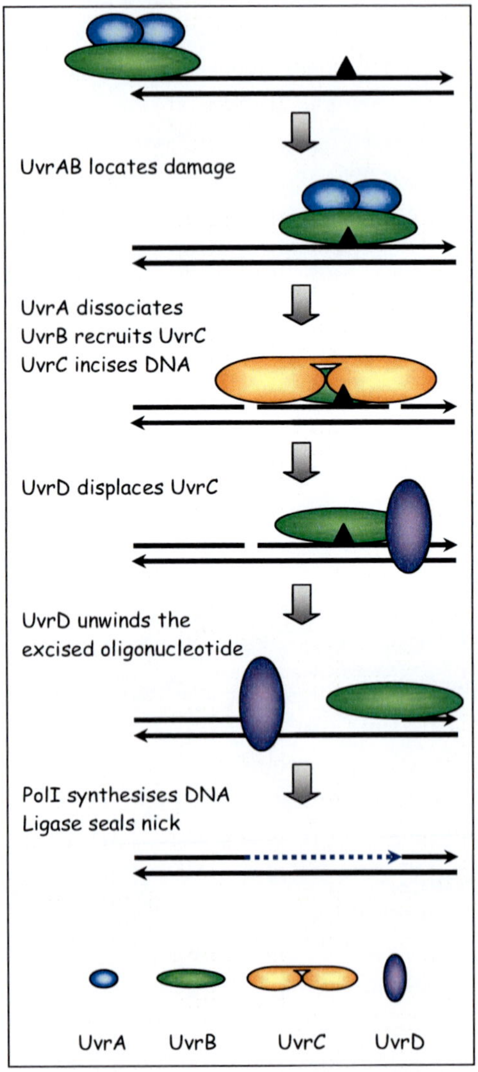

Figure 1. A schematic of the processes involved in nucleotide excision repair. The UvrAB complex scans DNA and identifies damaged nucleotides, following which UvrA dissociates leading to a stable UvrB-DNA pre-incision complex. UvrC binds to this UvrB-DNA complex and is responsible for incision of the DNA, first 4 or 5 bases 3', and then 8 bases 5', to the damage. The helicase UvrD then displaces UvrC and unwinds the excised oligonucleotide containing the lesion. DNA polymerase I fills in the gap and displaces UvrB, and finally the nick is sealed by DNA ligase.

3.2. Cho (UvrC homologue)

More recently, an additional protein with some sequence similarity to UvrC has been identified in *E. coli* and termed Cho (*uvrC* homologue). It has been demonstrated that this protein, in the presence of UvrA and UvrB, is able to catalyse the incision 3' to the lesion but not that on the 5' side. Repair is, therefore, effected by the combined actions of Cho and UvrC, which makes the 5' incision, and Cho appears to play a particular role in the repair of bulky adducts to DNA (Moolenaar *et al.*, 2002). A second protein with similarity to UvrC also exists in *M. tuberculosis* (Rv2191) although in this case there is also an additional domain that shows some sequence homology to the proofreading exonuclease domain of DNA polymerase III. For this reason the possibility of an alternative mechanism of repair involving a single incision followed by exonuclease digestion has been raised (Moolenaar *et al.*, 2002; Van Houten *et al.*, 2002), but this hypothesis has yet to be addressed experimentally. Interestingly, the sequence of this protein is conserved throughout its length in each of *M. bovis*, *M. marinum* and *M. avium* subsp. *paratuberculosis*, although the corresponding locus in *M. leprae* is a pseudogene. In *M. tuberculosis*, Rv2191 is upregulated following DNA damage (Rand *et al.*, 2003) as well as in both naïve and activated mouse macrophages from both wild-type and iNOS$^{-/-}$ mice (Schnappinger *et al.*, 2003), suggesting that Rv2191 may be of some importance for *in vivo* survival of *M. tuberculosis*.

3.3. UvrD: A Helicase

As discussed previously, UvrD acts in conjunction with the UvrABC complex in NER, and it has been shown that other *E. coli* DNA helicases cannot substitute for this function of UvrD, suggesting that its activity is specific to this gene (Mendonca *et al.*, 1995; Mendonca *et al.*, 1993). In *E. coli* UvrD also functions in mismatch repair (see later) and has been proposed to play a role in recombination (Mendonca and Matson, 1995) and replication (Moolenaar *et al.*, 2000). In support of the latter, mutation of any one of UvrA, UvrB or UvrD in strains lacking DNA polymerase I has been shown to be lethal in *E. coli* (Moolenaar *et al.*, 2000), suggesting an essential role for these proteins in an alternative replication pathway. Furthermore, both *E. coli* and *M. tuberculosis* UvrDs have been found to be active on forked DNA substrates designed to represent replication intermediates suggesting a role for UvrD in replication fork reactivation (Cadman *et al.*, 2006; Curti *et al.*, 2007). Interestingly, *M. tuberculosis* possesses two *uvrD* homologues (Rv0949 or *uvrD1*, and Rv3198c or *uvrD2*) as well as a further two genes which each contain a UvrD helicase domain (Rv3201c and Rv3202c). All four of these genes are conserved in *M. bovis*, *M. marinum* and *M. avium* subsp. *paratuberculosis* and the first two are also intact in *M. leprae* while the latter two are present as pseudogenes in this species. The expression of all four of these genes is increased in *M. tuberculosis* following DNA damage (Rand *et al.*, 2003), with a more marked effect seen for Rv3201c and Rv3202c. Rv3202c is also induced in activated mouse macrophages and in response to H_2O_2 (Schnappinger *et al.*, 2003).

3.4. Other NER Genes: *mfd* and *ercc3*

The *mfd* gene is a widely conserved bacterial protein that couples DNA repair with transcription. Mfd recognises RNA polymerase stalled at a non-coding template site of DNA damage, disrupts the transcription complex releasing the transcript and enzyme, and recruits UvrABC to the damaged site (Roberts and Park, 2004). An *mfd* homologue has been identified in *M. tuberculosis* (Cole *et al.*, 1998) and is also present in *M. bovis*, *M. marinum*, *M. avium* subsp. *paratuberculosis* and *M. leprae* suggesting it is important in mycobacterial survival. The *mfd* gene has recently been shown to be upregulated in *M. tuberculosis* in infected mouse and human macrophages (Cappelli *et al.*, 2006; Rachman *et al.*, 2006a) as well as following infection of SCID mice (Talaat *et al.*, 2004).

Discovered as a result of the human disorder Xeroderma pigmentosum, *ercc3* (excision repair cross-complementing subunit III) (Bootsma *et al.*, 1995) was only later identified in bacteria. The first prokaryotic *ercc3* gene was identified in *M. leprae* (Poterszman *et al.*, 1997), and homologues are also present in *M. tuberculosis*, *M. bovis*, *M. marinum*, and *M. avium* subsp. *paratuberculosis*. The conservation across mycobacterial species suggest that the protein encoded by *ercc3* plays an important function, although mutational studies to examine its role have yet to be described. In eukaryotes the ERCC3 protein functions as part of a multiprotein complex involved in NER and transcription; it has 3'-5' helicase activity and acts by opening DNA either around the RNA transcription start site or the site of DNA damage (Reardon and Sancar, 2005).

4. Base Excision Repair (BER)

Base excision repair is responsible for correcting most of the DNA damage caused by exposure to reactive oxygen intermediates and other metabolites that oxidise and alkylate DNA (Demple and Harrison, 1994; Gros *et al.*, 2002). This pathway is a two stage process. The key enzymes involved in the first step are DNA glycosylases, which remove specific damaged or mispaired bases by catalysing the hydrolysis of N-glycosylic bonds linking particular types of chemically altered bases to the sugar-phosphate backbone. This results in the formation of apurinic or apyrimidinic (AP) sites, which are further processed in the second step of repair by AP endonucleases, which hydrolyse the 5' phosphodiester bond immediately adjacent to the AP site to create a single nucleotide gap in one strand. This provides a substrate for non-repair specific enzymes (deoxyribophosphodiesterase, polymerase and ligase) to process and fill in.

There are two kinds of DNA glycosylases: monofunctional and bifunctional. The former cleave the N-glycosylic bond and release the modified base, leaving behind an AP site for recognition and cleavage by an AP endouclease (Krokan *et al.*, 1997). The bifunctional glycosylases similarly remove the modified base but in a concerted manner also cleave the phosphodiester bond 3' to the resulting AP site by a β- or β,δ-elimination mechanism (β-lyase activity) generating a single-strand break with 5'-phosphate and 3'-phosphoglycolate (PGA) or 3'-phosphate extremities respectively. As a 3'-terminal sugar derivative or 3'-phosphate cannot be extended by DNA polymerases further cleavage 5' to the site by an AP endouclease is required before repair can be completed via DNA polymerase and DNA ligase (Gros *et al.*, 2002) (Figure 2).

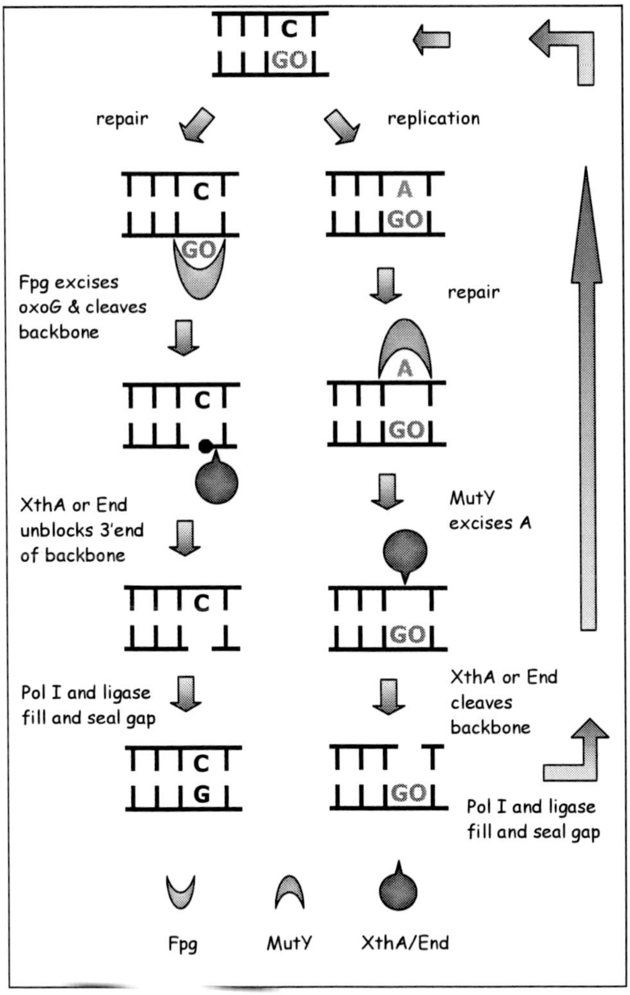

Figure 2. A schematic of the processes involved in the repair of 8-oxo-7,8-dihydroguanine by base excision repair, comparing the bifunctional DNA glycosylase Fpg (MutM) with the monofunctional DNA glycosylase MutY. Oxidised guanine (GO) opposite a C is repaired (left-hand side figure) when it is recognized by Fpg, which forms a complex in which the damaged base is extruded from the DNA helix. Fpg removes the damaged base and cleaves the backbone to leave a blocked 3'-terminus. Further processing by an AP endonuclease, XthA or End, removes this block and permits fill-in synthesis by DNA polymerase I followed by ligation to seal the nick. Alternatively, replication may occur before repair in which case A may be inserted opposite GO (right-hand side figure). In this case, the mis-paired A is recognized and removed by MutY, but the backbone remains intact. The backbone at the resulting abasic site is cleaved by XthA or End, again permitting polymerase I and ligase to fill-in and seal the gap.

4.1. Bifunctional DNA Glycosylases: *fpg*, *nei* and *nth*

One of the most common and stable oxidation products of guanine is 8-oxo-G (8-oxo-7,8-dihydroguanine) formed when the ring atom of the purine undergoes oxidation (Demple and Harrison, 1994). The Fpg protein (formamidopyrimidine-DNA glycosylase, also known as MutM) recognises these oxidised residues in the DNA, and incises the DNA at the damaged base (Friedberg *et al.*, 1995). If un-repaired, adenine (A) can pair opposite 8-oxo-G in subsequent DNA replication causing G·C-T·A transversions. To counteract this, if the 8-oxo-G (GO) lesion is not removed prior to replication, excision of A from these mispairs by MutY is initiated (see monfunctional DNA glycosylases and Figure 2). The Fpg protein also recognises and excises other damaged purine products such as 2,6-diamino-4-hydroxy-5-*N*-methylformamidopyrimidine (Fapy) lesions, and some pyrimidine oxidation products including 5-hydroxycytosine and 5-hydroxyuracil (Gros *et al.*, 2002). Fpg has been shown to be involved in the repair of DNA lesions induced by hydrogen peroxide in *E. coli* (Asad *et al.*, 1995).

A number of *fpg* gene homologues have been identified in *M. tuberculosis* (Cole *et al.*, 1998): Rv2924c which is the gene annotated as *fpg*, Rv0944 which is described as a possible Fapy DNA glycosylase with high sequence homology to *fpg*, and Rv2464c which has been identified as a possible *fpg* DNA glycosylase (http://genolist.pasteur.fr/TubercuList/). All three of these homologues are conserved in *M. bovis*, *M. marinum* and *M. avium* subsp. *paratuberculosis*, whilst only the homologue of Rv2924c is found as a functional gene in the reduced genome of *M. leprae*, with the other two orfs being represented by pseudogenes (Table 1). In *M. tuberculosis* expression of Rv2924c may show a small increase following infection of mouse macrophages (Schnappinger *et al.*, 2003) and Rv2464c has been reported to be expressed at an elevated level in clinical lung samples (Rachman *et al.*, 2006b). These data coupled with the potential functional redundancy suggest an important role for *fpg* activity in *M. tuberculosis* during intracellular survival.

The Nth protein (endonuclease III) is a DNA glycosylase that has broad substrate specificity; it can excise ring-saturated, ring-opened and ring-fragmented pyrimidines, for example thymine glycol, urea and dihydro-uracils (Friedberg *et al.*, 1995). A *nth* gene, Rv3674c, has been identified in *M. tuberculosis* (Cole *et al.*, 1998) and homologues are found in *M. bovis*, *M. marinum*, *M. avium* subsp. *paratuberculosis* and *M. leprae*. The Nei protein (endonuclease VIII) was discovered after an *E. coli* mutant defective in *nth* activity was still able to excise thymine glycol from damaged DNA (Melamede *et al.*, 1994). Thus, its activity is very similar to that of Nth, although it has broader substrate specificity, as it has been shown to be involved in repair of lesions such as 5-hydroxycytosine and 5-hydroxyuracil (Dizdaroglu *et al.*, 2000; Jiang *et al.*, 1997b), which may be related to its similarity to Fpg (Jiang *et al.*, 1997a). Although Nei has been suggested to serve as a back-up repair system for removal of 8-oxo-G lesions in the absence of Fpg and MutY (Blaisdell *et al.*, 1999), more recent reports suggest that this is not the case (Wiederholt *et al.*, 2005). The *nei* gene, Rv3297, has been identified in *M. tuberculosis* (Cole *et al.*, 1998), and although it is conserved in *M. bovis*, *M. marinum* and *M. avium* subsp. *paratuberculosis* it is absent from *M. leprae*. There is currently no information available about the role of Nth or Nei in mycobacteria, but interestingly an *nth nei* double mutant in another intracellular pathogen, *S.*

typhimurium, showed a 5-fold attenuation in growth in a mouse model of infection as compared to wild-type (Suvarnapunya *et al.*, 2003).

4.2. Monofunctional DNA Glycosylases: *mutY, ung, alkA, tagA*

The MutY protein is primarily an adenine DNA glycosylase that can initiate base excision repair of A opposite 8-oxo-G, G or C. Following this the excised A nucleotide can be replaced by accurate repair synthesis, resulting in correction of the lesion or reformation of 8-oxo-G·C thus allowing another chance for Fpg-mediated repair (Friedberg *et al.*, 1995). Although originally classified as a bifunctional DNA glycosylase, subsequent studies suggested that MutY lacks AP lyase activity (Zharkov and Grollman, 1998). Interestingly, it was found that MutY forms a long-lived complex with its DNA substrate which it was proposed blocks access of Fpg to oxo-G and so prevents the formation of a double-stranded break. The *mutY* gene in *M. tuberculosis* is Rv3589, and homologues are also present in *M. bovis*, *M. marinum*, *M. avium* subsp. *paratuberculosis* and *M. leprae*. An *M. tuberculosis* strain deficient for *mutY* appears to have heightened susceptibility to H_2O_2 (Rand, 2003), suggesting it is important in the repair of oxidative damage.

Uracil DNA glycosylase (*ung*) is a highly conserved protein present from prokaryotes to humans (Krokan *et al.*, 1997), and is specific to repair of uracil arising in DNA either as a consequence of DNA damage, such as deamination, or of misincorporation. The gene encoding Ung in *M. tuberculosis* is Rv2976c, and again this gene is conserved in *M. bovis*, *M. marinum*, *M. avium* subsp. *paratuberculosis* and *M. leprae*. In contrast, these species of mycobacteria lack a homologue of the mismatch-specific uracil DNA glycosylase encoded by the gene *mug*, the role of which *in vivo* is not clear although Mug has been shown to be important for processing ethenobases (Jurado *et al.*, 2004) and appears to contribute to mutation avoidance in stationary phase (Mokkapati *et al.*, 2001). Abrogation of Ung activity in *M. smegmatis* confers an increased mutator phenotype and sensitivity to reactive nitrogen intermediates generated by acidified nitrite (Venkatesh *et al.*, 2003). Interestingly, *M. smegmatis* defective for *ung* exhibits significantly reduced survival in a mouse macrophage infection model with a nearly total abolition of multiplication and a much-decreased residency in macrophages stimulated with interferon γ (Venkatesh *et al.*, 2003). In *M. tuberculosis*, the *ung* gene has been predicted to be required for survival *in vivo* at early stages of infection in a mouse model using a transposon mutagenesis approach (Sassetti and Rubin, 2003).

The *alkA* and *tagA* DNA glycosylases mainly recognise alkylated bases; *tagA* is specific for 3-methylpurines, while *alkA* removes both 3-methyladenine and hypoxanthine, the latter arising from the deamination of adenine, a process that is stimulated by nitric oxide (Gros *et al.*, 2002). A *tagA* gene is present in *M. tuberculosis* (Rv1210), *M. bovis*, *M. marinum*, *M. avium* subsp. *paratuberculosis* and *M. leprae*, but studies on its role have not been reported. A homologue of *alkA* is found in *M. tuberculosis* (Rv1317c), but as well as a region of homology to AlkA, the protein encoded by this gene also contains an adenosine deaminase domain having similarity to the methyl acceptor and regulatory parts of the *E. coli* Ada protein; the role of this part of the protein in the absence of the methyl transferase domains is unknown. Although this arrangement is conserved in *M. marinum* and *M. avium* subsp.

paratuberculosis, in *M. leprae alkA* is present as a pseudogene, while in *M. bovis* a single base transition introduces a premature stop codon, splitting the corresponding locus into two open reading frames. Interestingly, this disruption occurs within the Ada adenosine deaminase domain, between the metal binding subdomain and the AraC regulatory subdomain, whilst leaving the AlkA domain intact; whether the two fragments of the protein interact to restore the Ada domain, mutation at this site is inconsequential, or this change affects the function of the protein is not known.

The *alkA* gene in *M. tuberculosis* is highly upregulated in response to H_2O_2 (Schnappinger *et al.*, 2003). However, although an *M. tuberculosis* strain defective in *alkA* and *ogt* (an alkylguanine-DNA alkyltransferase involved in direct reversal of DNA damage) was hypersensitive to the alkylating agent MNNG (*N*-methyl-*N'*-nitro-*N*-nitrosoguanidine), it showed no phenotype *in vivo* using a mouse infection model (Durbach *et al.*, 2003). Intriguingly, an additional gene similar to several eukaryotic 3-methyladenine DNA glycosylases, annotated as *mpg*, is also present in *M. tuberculosis* (Rv1688), *M. bovis*, *M. marinum*, *M. avium* subsp. *paratuberculosis* and *M. leprae*. The conservation of this gene across these species suggests that it plays a significant role. Although functional studies are awaited, the expression of *mpg* in *M. tuberculosis* has been reported to be elevated in clinical lung samples (Rachman *et al.*, 2006b).

4.3. AP Endonucleases: xthA and end (nfo)

The AP sites generated by removal of the damaged base require further biochemical repair events via *apurinic/apyrimidinic* (AP) endonucleases which catalyse the incision of the DNA backbone exclusively at AP sites, thus preparing the DNA for repair synthesis and DNA ligation. The two genes identified in *E. coli* that function here are *nfo* (endonuclease IV) and *xthA* (exonuclease III). Both the Nfo and XthA proteins catalyse the formation of single-stranded breaks at sites of base loss in duplex DNA, via attack on the phosphodiester bond 5' to the site of base loss, leaving 3'OH groups. The only distinguishing difference between the two proteins is the presence of a 3'-exonuclease activity in XthA (Friedberg *et al.*, 1995). In *E. coli,* XthA makes up the majority of AP endonuclease activity, comprising approximately 90% (Ljungquist *et al.*, 1976). The importance of these processes for the repair of oxidative damage is indicated by the observation that in *E. coli* mutants of *xthA* are hypersensitive to H_2O_2, and *nfo/xthA* double mutants are even more sensitive (Galhardo *et al.*, 2000; Zhang *et al.*, 1992). In addition, strains of the intracellular pathogen *Salmonella typhimurium* with mutations in *xthA/nfo* are attenuated in mice (Suvarnapunya *et al.*, 2003). Homologues of both these genes have been identified in *M. tuberculosis*. The homologue of *nfo* (Rv0670) has been termed *end* (Cole *et al.*, 1998) and is conserved in *M. bovis*, *M. marinum*, *M. avium* subsp. *paratuberculosis* and *M. leprae*, while *xthA* (Rv0427c in *M. tuberculosis*) is only found as a pseudogene in *M. leprae*. Both the *xthA* and *end* genes of *M. tuberculosis* have been predicted to be required for survival at early stages of infection *in vivo* using a mouse model (Sassetti and Rubin, 2003).

5. Alternative Excision Repair Pathways

5.1. Nucleotide Incision Repair (NIR)

Recently, a novel mechanism of DNA repair has been described and termed nucleotide incision repair (Ischenko and Saparbaev, 2002). In *E. coli* this pathway is initiated by Nfo, which nicks the DNA on the 5' side of the damaged base, generating a 3'-hydroxyl terminus suitable for DNA synthesis by polymerase I, and leaving the damaged nucleotide dangling at the 5' terminus where it can be removed by the exonuclease activity of PolI. This pathway is primarily targeted to oxidative damage and Nfo incises the DNA adjacent to a range of oxidatively damaged bases. In contrast to the role of Nfo in base excision repair, this repair activity of Nfo is independent of DNA glycosylases. Indeed, it has been shown that these two activities of Nfo are genetically separable (Ishchenko *et al.*, 2006). Individual mutations of three amino acids each resulted in Nfo proteins that retained AP endonuclease and 3'-phosphodiesterase activities but were deficient in NIR. Interestingly, strains of *E. coli* in which the only AP endonuclease was one of these mutant Nfos exhibited heightened sensitivity to the oxidative stress agent t-butylhydroperoxide similar to strains completely lacking AP endonucleases, but were complemented for resistance to the alkylating agent MMS that is dependent on removal by a DNA glycosylase and repair via the BER pathway. These three critical amino acids are conserved in the proteins encoded by the *end* genes of *M. tuberculosis*, *M. bovis*, *M. marinum*, *M. avium* subsp. *paratuberculosis* and *M. leprae*, suggesting that this pathway may also be functional in mycobacteria.

5.2. Excision Repair by Endonuclease V

Another alternative excision repair pathway has been defined in *E. coli* that is particularly important for the removal of xanthine from DNA. Xanthine is formed as a consequence of deamination of guanine and is highly mutagenic, leading predominantly to G.C to A.T transitions. However, no glycosylase that recognizes xanthine has been identified in *E. coli*. Instead repair is effected via incision of the second phosphodiester bond 3' to the lesion by endonuclease V, encoded by the gene *nfi* (Kow, 2002). This is presumably followed by removal of the damage and resynthesis but the details of how this occurs are currently not known. This pathway does not appear to be present in mycobacteria, as no homologue of Nfi is identifiable in any of *M. tuberculosis*, *M. bovis*, *M. marinum*, *M. avium* subsp. *paratuberculosis* or *M. leprae*.

6. Mismatch Repair

Replication errors are probably one of the most common sources of mismatched base pairs in DNA. Nucleotide misincorporation during DNA synthesis occurs at a frequency of $10^{-4} - 10^{-5}$ as a consequence of the nature of the DNA polymerase combined with the base pair geometry, although proof-reading exonuclease activities of polymerases can reduce the overall error rate to ~10^{-7} per bp per replication (Iyer *et al.*, 2006). Mismatches can also arise as a result of recombination between two DNAs that contain differences in their sequences, as well as via aberrant processing by DNA repair systems.

Systems to repair mismatches have been identified in both prokaryotes and eukaryotes and appear to be closely related. The best characterised system is that of *E. coli*. The key difference in the repair of mismatched bases as opposed to other forms of DNA repair is how to discriminate between the correct and the incorrect base, as both strands contain a normal nucleotide at the mismatch. In *E. coli* this is done by linking the base to be excised to the methylation status of adenine in GATC sequences on the same DNA strand. Methylation of DNA by the sequence-specific *dam* methylase lags behind DNA replication, so a newly synthesised DNA strand is undermethylated compared to the parental strand (Lobner-Olesen *et al.*, 2005). For this reason the system is also known as the methyl-directed mismatch repair system.

The enzymes whose functions are specific to mismatch repair are MutS, MutL and MutH, while helicase II (UvrD) also plays an important role in the process. MutS plays the key role of recognizing and binding to the mismatch, and in the presence of ATP forms loop structures. MutL binds to this protein-DNA complex in a process dependent on ATP but not on ATP hydrolysis, and it has been proposed that MutL accelerates the translocation of the complex along the DNA until a hemi-methylated GATC site bound by MutH is reached (Allen *et al.*, 1997). MutH is the endonuclease that cleaves the unmethylated strand just 5' to the GATC sequence when activated by the MutS-MutL-DNA complex, and thus confers strand specificity to the whole process. Following incision, UvrD is loaded onto the nicked site in a process requiring interaction with MutL (Mechanic *et al.*, 2000), and unwinds the incised strand preferentially in the direction of the mismatch (Dao and Modrich, 1998). The released single strand is degraded by exonucleases to just beyond the mismatch, allowing resynthesis by PolIII to occur. The incised GATC may be located either 5' or 3' to the mismatch on the unmethylated DNA strand. When the incision is on the 5' side of the mismatch 5'-3' exonuclease function is provided by either ExoVII or by RecJ. When the incision is 3' to the mismatch 3'-5' exonuclease activity is supplied by ExoI, ExoVII or ExoX (Figure 3).

The use of methylation status to discriminate between parental and daughter strands is not highly conserved. However, the requirement for a hemi-methylated site to confer strand specificity can be circumvented even in *E. coli* by the presence of a persistent strand break (Langle-Rouault *et al.*, 1987). Thus, strand breaks in a daughter strand such as would arise by the discontinuous synthesis of the lagging strand could be sufficient to confer discrimination, and this is thought to be the case in the Hex-dependent mismatch repair system of *Streptococcus pneumoniae* (Claverys and Lacks, 1986). Despite this the HexA and HexB proteins of *S. pneumoniae* are homologous to MutS and MutL respectively (Priebe *et al.*, 1988; Prudhomme *et al.*, 1989). Indeed, MutS and MutL homologues are highly conserved and widely distributed (Fishel, 1998). In *E. coli*, genetic inactivation of the mismatch repair system by mutation of any of *mutS*, *mutL*, *mutH* or *uvrD* leads to elevated mutation rates, particularly base substitutions and frameshifts (Claverys and Lacks, 1986). In addition, such a defect increases by several orders of magnitude the frequency of recombination between quasi-homologous, often termed homeologous, sequences which contain 10-12% heterology.

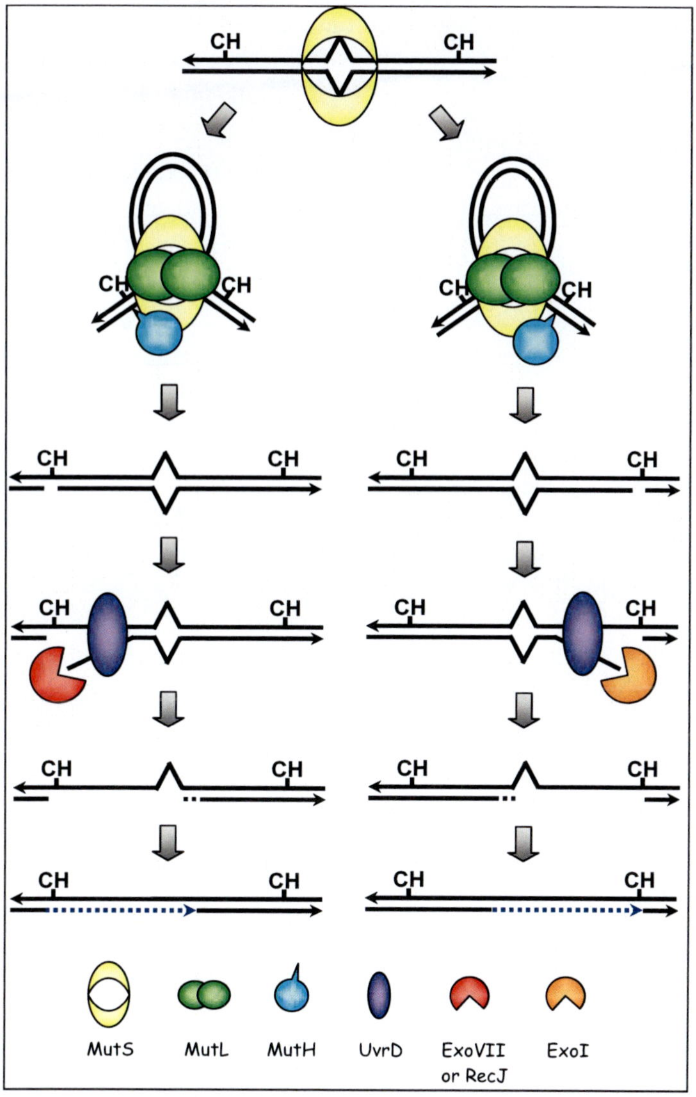

Figure 3. A schematic of the processes involved in mismatch repair in *E. coli*. This pathway is not present in mycobacteria. The mismatch is recognized by MutS, and MutL binds to the MutS-DNA complex to facilitate translocation to a hemi-methylated GATC site bound by MutH. MutH cleaves the unmethylated strand when activated by the MutS-MutL-DNA complex. The helicase UvrD then unwinds the incised strand preferentially in the direction of the mismatch, and the released single strand is degraded by either a 5'-3' exonuclease (ExoVII or RecJ) or a 3'-5' exonuclease (ExoI, ExoVII or ExoX) to just beyond the mismatch. Following this, resynthesis by PolIII can occur and finally the DNA sealed by ligase.

Perhaps surprisingly, analysis of the *M. tuberculosis* genome sequence suggested that this species lacks a system for mismatch repair (Mizrahi and Andersen, 1998). None of the key components of mismatch repair - MutS, MutL or MutH - could be identified by homology in *M. tuberculosis*, and neither are there any recognisable orthologues in any of *M. bovis*, *M. marinum*, *M. avium* subsp. *paratuberculosis*, or *M. leprae*. In addition, no homologues of the exonucleases RecJ or exonuclease I (*sbcB* or *xonA*) are identifiable in any of these species, although the subunits of exonuclease VII (*xseA* and *xseB*) are conserved in them all. Furthermore, the lack of a classical mismatch repair system in mycobacteria has been demonstrated experimentally (Springer *et al.*, 2004). In both *M. smegmatis* and *M. bovis* BCG frameshift mutations occurred at a high rate relative to the general mutation rate, although the latter was not exceptionally high. It was also found that flanking homeologous sequences did not reduce the rate of recombination if strand transfer was first initiated in a region of perfect sequence identity. These two phenotypes are consistent with the lack of a functional mismatch repair system. Nevertheless, recombination did not occur between sequences that were around 11% divergent throughout, indicating that mycobacteria must possess other ways to control homeologous recombination, possibly involving a stringent requirement for sequence homology at the level of RecA-mediated strand transfer.

7. Recombination Repair

Recombination repair is complex and requires two DNA duplexes as well as a large number of proteins. Although there are multiple pathways of recombination repair, virtually all of those characterized rely on RecA, a protein that plays a central role in the process of recombination (Kowalczykowski *et al.*, 1994). Types of repair described include *recA*-dependent excision repair, postreplication repair (daughter strand gaps) and repair of double-stranded breaks (Smith, 2004) (Figure 4). The two main recombination repair pathways identified from studies in *E. coli* are the RecBCD pathway and the RecFOR pathway (Kowalczykowski *et al.*, 1994). Both require RecA, but the latter also utilises RecJ (a 5' → 3' exonuclease) and RecQ (a helicase). Whilst the Rec proteins that are common to both *E. coli* and *M. tuberculosis* show a high degree of sequence homology (Cole *et al.*, 1998), certain key components of *E. coli* pathways are missing; in particular there are no clear homologues of *recJ* or *recQ* (Mizrahi and Andersen, 1998; Muniyappa *et al.*, 2000).

7.1. RecA

The RecA protein has an important role in homologous recombination and recombination repair by means of its activity as a recombination exchange protein, whereby it promotes homologous pairing and strand-exchange in an ATP-dependent reaction (Cox, 1991; Kowalczykowski *et al.*, 1994). It is also central to the control of the SOS response (see below) and plays a role in replication following fork arrest (Courcelle and Hanawalt, 2003). Strains of *E. coli* defective in *recA* have been shown to exhibit a severe decrease in recombination efficiency, and in addition are sensitive to DNA damage, affected in SOS mutagenesis, and exhibit aberrant cell division and chromosome segregation (Clark, 1973; Kowalczykowski *et al.*, 1994; Lloyd *et al.*, 1974).

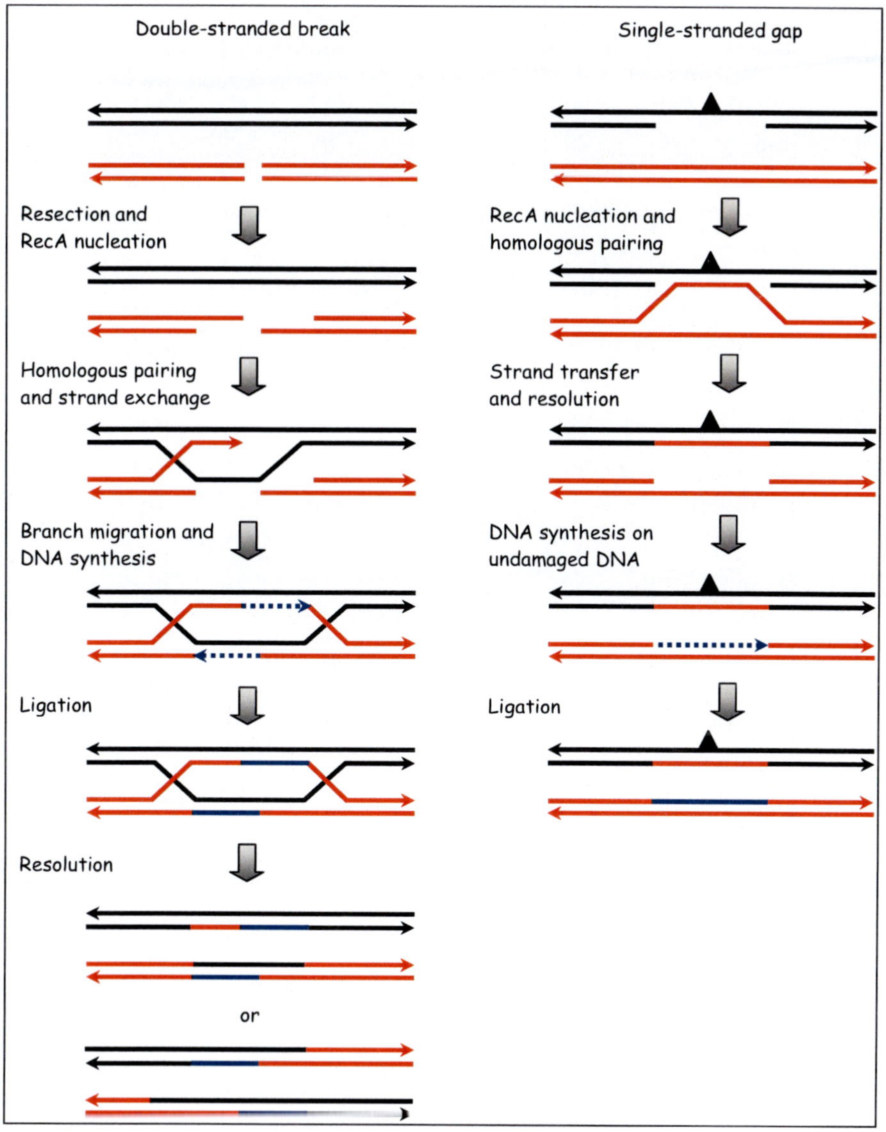

Figure 4. A simplified schematic of recombinational repair processes at a double-stranded DNA break (left-hand side) and at a single-stranded DNA gap (right-hand side). The initial step is the formation of a RecA nucleo-protein filament; this is usually facilitated by RecBCD at dsDNA ends or RecFOR at ssDNA gaps, although other combinations are possible. RecA then catalyses homologous pairing and strand exchange. Following this, branch migration is mediated by RuvAB or RecG, concomitantly with DNA synthesis by PolI and ligation, and finally resolution of the recombination intermediates is accomplished by the action of RuvC.

Not surprisingly in view of its central role in bacterial DNA metabolism, a *recA* gene is present in all the species of mycobacteria so far examined. However, the *recA* loci of certain mycobacteria differ from those of most species and other bacteria in that they contain large intervening sequences termed protein introns or inteins that are excised by an unusual protein-splicing reaction (Davis *et al.*, 1992; Saves *et al.*, 2000). Nevertheless, the *M. tuberculosis recA* gene has been shown to complement an *E. coli recA* mutant to varying extents for functions in recombination, DNA repair, mutagenesis and SOS induction (Davis *et al.*, 1991). Furthermore, the presence of the intein does not interfere with the ability of RecA to function in homologous recombination (Frischkorn *et al.*, 1998; Papavinasasundaram *et al.*, 1998). Biochemical analyses have revealed that the *M. tuberculosis* RecA protein differs from the *E. coli* enzyme in exhibiting a more pronounced pH dependence of strand exchange as well as reduced efficiency of ATP binding and catalysis (Kumar *et al.*, 1996; Vaze and Muniyappa, 1999), the latter observation being explained by a wider binding cleft observed in the crystal structure (Datta *et al.*, 2000; Datta *et al.*, 2003).

In the non-pathogenic *M. smegmatis*, a *recA* deletion strain exhibited enhanced sensitivity to UV irradiation and failed to undergo homologous recombination (Papavinasasundaram *et al.*, 1998). Deletion of *recA* in *M. bovis* BCG, whilst conferring sensitivity to DNA damaging agents, did not affect the ability of the bacteria to survive in an *in vitro* model of dormancy, although it did result in heightened sensitivity to metronidazole under these conditions. Perhaps surprisingly, the mutant strain also was unaffected in its ability to grow and persist in a mouse infection model (Sander *et al.*, 2001). A strain of *M. tuberculosis* deleted for *recA* has been reported (Davis *et al.*, 2002b), but so far characterization of this strain has been limited to studies of gene regulation (Davis *et al.*, 2002b; Rand *et al.*, 2003). However, expression of the *M. tuberculosis recA* gene has been shown to be induced in response to H_2O_2 (Schnappinger *et al.*, 2003) as well as to DNA damage by mitomycin C or ofloxacin (Movahedzadeh *et al.*, 1997a; Papavinasasundaram *et al.*, 2001).

7.2. RecX

It is only recently that the role of RecX in recombination has begun to be understood. The *recX* gene was originally identified in *Pseudomonas aeruginosa* as it was required for survival when RecA was expressed at high levels (Sano, 1993). This suggested that RecX played some regulatory role related to RecA function but did not indicate whether this effect was exerted at the level of control of expression or of activity. In *P. aeruginosa recX* was located immediately 3' to *recA*, and a homologous gene was subsequently identified in the corresponding location in other bacterial species including *M. leprae*, as well as in *E. coli* (De Mot *et al.*, 1994). In *M. leprae* the 5' end of *recX* actually overlaps the 3' end of *recA*, and the same arrangement also occurs in *M. tuberculosis*, *M. bovis*, *M. marinum* and *M. smegmatis* (Papavinasasundaram *et al.*, 1997). The requirement for RecX when RecA is expressed at high level was also demonstrated in mycobacteria (Papavinasasundaram *et al.*, 1998).

The first clue that RecX might act on the RecA protein came from the observation that the levels of *recA* transcription did not appear to be altered in a *recX* mutant of *Streptomyces lividans* (Vierling *et al.*, 2000) *or Xanthomonas oryzae pv. oryzae* (Sukchawalit *et al.*, 2001). Direct protein-protein interactions between RecX and RecA were first demonstrated for the *M. tuberculosis* proteins, both *in vitro* by affinity chromatography and surface plasmon resonance and *in vivo* by immunoprecipitation (Venkatesh *et al.*, 2002). Although interaction with RecX did not affect the ability of RecA to bind to ssDNA or ATP, it did inhibit its DNA-dependent ATPase activity and also strand exchange *in vitro* (Venkatesh *et al.*, 2002). Mycobacterial RecX also impedes RecA-stimulated LexA cleavage *in vitro* (S. Manglani, K.G. Papavinasasundaram, B. Gopal and E.O.D., unpublished).

These *in vitro* properties of RecX appear to be widespread with similar effects being described in *E. coli* (Stohl *et al.*, 2003) and *Deinococcus radiodurans* (Sheng *et al.*, 2005). In these organisms *recX* mutant strains have also been studied, revealing that both overexpression and loss of RecX can affect phenotypes such as uv survival and recombination. In *E. coli* overexpression of RecX *in vivo* inhibited the coprotease activity of RecA, and so induction of the SOS response, as well as recombination assessed by P1 transduction, but no differences in these properties were detectable when comparing the strain lacking *recX* with the wild-type (Stohl *et al.*, 2003). Thus, it was suggested that RecX has a role during the SOS response. It has been reported that in *E. coli* the *recX* gene is expressed at a much lower level that *recA* owing to the presence of a transcriptional terminator between the two genes (Pages *et al.*, 2003), but it would appear that this is unlikely in mycobacteria where the two genes overlap.

The molecular basis for the effects of RecX on RecA function have been defined by detailed studies in *E. coli*. These have revealed that RecX inhibits the extension of RecA nucleoprotein filaments (Drees *et al.*, 2004). Owing to the unidirectional extension of such filaments this leads to net disassembly. This mechanism can explain why some RecA activities, such as strand exchange which requires intact RecA filaments, are inhibited at relatively low RecX concentrations while a higher amount of RecX relative to RecA is needed to inhibit others, such as LexA cleavage for which only short segments of RecA filament are necessary. Thus, RecX appears to act by affecting filament formation rather that the function of RecA within filaments. Structural analyses with higher relative concentrations of RecX have indicated that RecX binds deep within the major helical groove of a RecA filament where it would form a block to stand exchange (VanLoock *et al.*, 2003), raising the possibility of an alternative mode of action for RecX under some conditions. Intriguingly, the expression of the *recX* gene in *M. tuberculosis* is reported to be enhanced intra-cellularly in mouse macrophages (Schnappinger *et al.*, 2003) and in SCID mice (Talaat *et al.*, 2004).

7.3. RecBCD

The different recombination pathways are named by the proteins involved in the formation of a RecA nucleoprotein filament, which is responsible for the recombination process itself. In *E. coli* the first recombination pathway discovered was the RecBCD pathway which is particularly important in the repair of double-stranded breaks (Kowalczykowski *et al.*, 1994). The RecBCD protein complex possesses both helicase and nuclease activities, and its primary role is to process a blunt double-stranded DNA end to form a 3' single-stranded tail to which RecA can bind (Kowalczykowski, 2000). In order to achieve this the complex actually contains two helicases of opposite polarity, RecB is a 3' – 5' helicase and RecD functions as a 5' – 3' helicase (Dillingham *et al.*, 2003; Taylor and Smith, 2003), that act on each strand of the DNA to confer high processivity. Initially, unwinding is coupled to degradation of the 3' terminated single strand of DNA (Dixon and Kowalczykowski, 1993). However, following interaction with a Chi site, a specific DNA sequence that is recognised as single-stranded DNA, RecBCD activity is modified such that the 5' single-stranded DNA is preferentially degraded, yielding a 3' single-stranded tail (Anderson and Kowalczykowski, 1997a). Concomitant with this change in nucleolytic activity RecBCD plays an active role in loading RecA onto the Chi-terminated 3' single-stranded DNA produced (Anderson and Kowalczykowski, 1997b) (Figure 5).

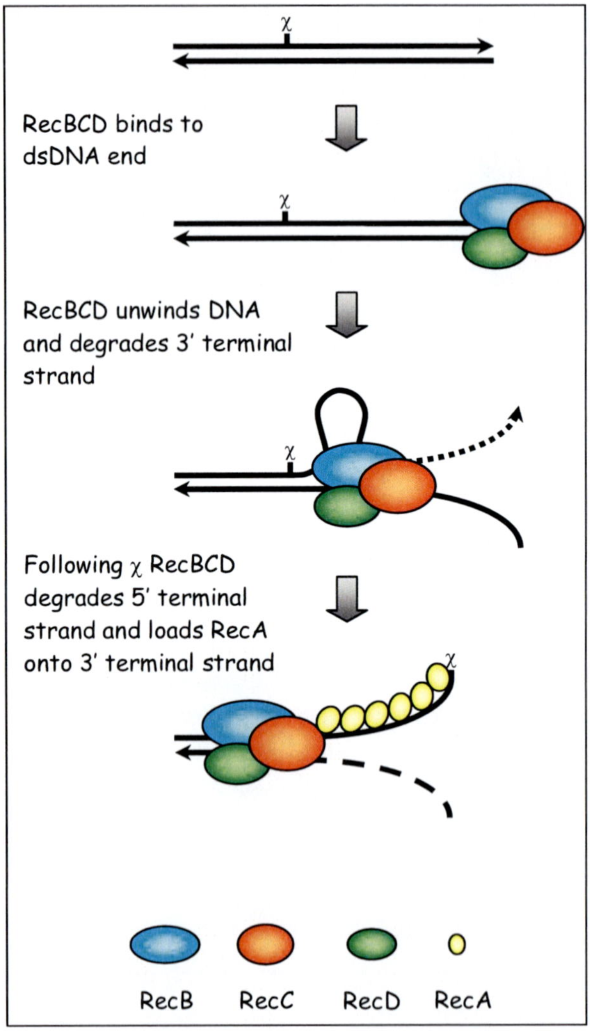

Figure 5. A schematic of RecBCD processing of a double-stranded DNA break. The RecBCD complex binds to dsDNA ends and unwinds the DNA, simultaneously degrading the 3'-terminal strand. Also, the differential speeds of the two helicases RecB and RecD results in a DNA loop in this strand. Following encounter with a Chi site, cleavage occurs preferentially on the 5'-terminal strand, albeit at a reduced frequency, leading to the formation of a single-stranded DNA with a 3'-terminal Chi site. RecBCD also plays a role in loading RecA onto this single-stranded DNA, thus forming the substrate for recombination.

In other Gram-positive bacteria studied, equivalent functions to those of the RecBCD complex are performed by the AddAB proteins, which are not closely related, and there are no sequence homologues of *recBCD* (Chedin and Kowalczykowski, 2002). However, RecBCD homologues are present in some species of mycobacteria, *M. tuberculosis* and *M. marinum*, although these genes are completely absent in *M. leprae*. Intriguingly, while *recC* and *recD* are also found in *M. bovis* and *M. avium* subsp. *paratuberculosis*, in each of these species the *recB* locus contains a separate frameshift splitting the region of homology into two open reading frames; whether the proteins encoded by these pairs of orfs are able to interact to restore function or these frameshifts result in inactivation of the RecBs is currently unknown.

Although the variable distribution of the *recBCD* genes amongst mycobacteria might suggest that they are not particularly important for survival, in another intracellular pathogen, *S. typhimurium*, *recBC* mutants, as well as *recA* mutants, are attenuated in mice, and sensitive to killing by reactive oxygen intermediates in macrophages *in vitro* (Buchmeier *et al.*, 1993). Interestingly, *recB*, but neither *recC* nor *recD*, has been reported to be expressed at elevated levels in clinical lung samples of TB infected patients (Rachman *et al.*, 2006b), while *recC* exhibited increased expression following infection of human macrophages (Cappelli *et al.*, 2006) or SCID mice (Talaat *et al.*, 2004). These observations imply that this pathway may have a role *in vivo* in *M. tuberculosis* as well.

7.4. RecFOR

The second recombination pathway in *E. coli* is the RecF or RecFOR pathway, which is primarily involved in initiating recombination at single-stranded gaps and is thought to play a significant role in the reactivation of stalled replication forks and the repair of single-stranded breaks (Morimatsu and Kowalczykowski, 2003). The RecFOR pathway can also act at dsDNA breaks in conjunction with the helicase RecQ and the exonuclease RecJ in *E. coli* (Kowalczykowski *et al.*, 1994), but these latter two proteins are absent from mycobacteria studied to date, as is helicase IV (encoded by *helD* in *E. coli*) which is also implicated in this pathway (Mendonca *et al.*, 1995). The precise functions of the RecFOR proteins are not completely clear, but it has been suggested that they are involved in modulating assembly and disassembly of RecA filaments (Bork *et al.*, 2001). Recently it has been shown that together they act to direct the loading of RecA onto gapped DNA that is coated with single-stranded binding (SSB) protein in a process that is dependent on the presence of a 5' base-paired junction as well as RecO-RecR and RecF-RecR interactions (Morimatsu and Kowalczykowski, 2003). It is thought that RecF or the RecFR complex recognises and binds to the ssDNA-dsDNA junction, then RecO binds to this complex and probably interacts with SSB. The RecFOR proteins have also been demonstrated to protect the nascent lagging strand of arrested replication forks (Chow and Courcelle, 2004). Another role from studies in *E.coli* has been suggested based on the observation that RecF interacts with RecX and protects RecA from the inhibitory effects of RecX during RecA filament formation (Lusetti *et al.*, 2006).

In contrast to the distribution of the *recBCD* genes, homologues of each of *recF*, *recO* and *recR* are identifiable in each of *M. tuberculosis*, *M. bovis*, *M. marinum*, *M. avium* subsp. *paratuberculosis* and *M. leprae*, perhaps suggesting that the RecFOR pathway is of greater importance to mycobacteria. Although no sequence homologues of *recQ* or *recJ* are present, it remains possible that equivalent functions could be provided by unrelated gene products, or these functions may not be required for the roles performed by RecFOR in mycobacteria. Little data is available on the respective functions of these proteins in *M. tuberculosis*, but *recF* has been predicted to be required for survival in mouse macrophages (Rangarajan *et al.*, 2002) supporting a potential role in pathogenesis.

7.5. RadA

The *radA* gene, named for its role in radiation resistance in *E. coli*, has also been reported to play a role in recombination repair processing double-stranded breaks (Song and Sargentini, 1996), but it should be noted that the bacterial RadA is quite different from the archaeal RadA that is a RecA protein homologue that catalyses strand exchange (Seitz *et al.*, 1998). The bacterial gene contains a putative zinc finger at its N terminus, shows some sequence similarity to RecA and DnaB (replicative DNA helicase) in the central region, and at the C-terminus is related to Lon protease (an ATP-dependent serine protease that binds DNA) although no protease activity has been detected (Beam *et al.*, 2002). The function of the bacterial RadA is largely unknown, but mutation of *radA* in *E. coli* has been reported to be synergistic with mutatons in *recG*, and to a lesser extent in *ruvA* or *ruvC*, suggesting a role in processing of branched DNA molecules or blocked replication forks (Beam *et al.*, 2002).

A *radA* homologue has been identified in *M. tuberculosis* from the genome sequence, and homologues are also found in *M. bovis*, *M. marinum* and *M. avium* subsp. *paratuberculosis*, while in *M. leprae* a pseudogene is present. In *M. tuberculosis* the *radA* gene is highly upregulated *in vitro* in response to H_2O_2 (Schnappinger *et al.*, 2003) and mitomycin C (Rand *et al.*, 2003). Moreover, its expression is reported to be enhanced in activated mouse macrophages (Schnappinger *et al.*, 2003) and clinical lung samples (Rachman *et al.*, 2006b), suggesting a role for RadA in protection against ROI and during infection.

7.6. RuvABC and RecG

Other recombination proteins have been identified in *E. coli* which work downstream of RecA function. These include RuvA, RuvB, RuvC and RecG which are involved in branch migration (RuvAB and RecG) and resolution (RuvC) of Holliday junctions (McGlynn and Lloyd, 2002; West, 1995), which can be formed at stalled replication forks as well as during recombination. RuvA binds to the Holliday junction as a tetramer to stabilise the square planar configuration of the junction, and facilitates the assembly of the hexameric RuvB helicase on opposite arms of the junction (Sharples *et al.*, 1999). The DNA is then 'pumped' out of the complex by translocation of RuvB resulting in branch migration. RuvC is thought to be recruited to the complex where it functions as a dimeric endonuclease to symmetrically cleave two of the four DNA strands, to achieve resolution of the Holliday junction (Dunderdale *et al.*, 1991). RecG is a structure-specific helicase which can also catalyse branch migration of Holliday junctions (Lloyd and Sharples, 1993), consistent with the finding that mutation of *recG* is synergistic with mutation of *ruvAB* (Lloyd, 1991). In addition, RecG promotes rescue of damaged forks by catalysing their unwinding and conversion to Holliday junctions (McGlynn and Lloyd, 2002).

Homologues for each of these genes (RuvA, RuvB, RuvC and RecG) have been identified in *M. tuberculosis* (Cole *et al.*, 1998), *M. bovis*, *M. marinum*, *M. avium* subsp. *paratuberculosis*, and *M. leprae*. This extensive conservation suggests that these genes provide an important function. The structure of the *M. leprae* RuvA protein has been determined in complex with a Holliday junction and revealed two tetramers bound either side of the junction (Roe *et al.*, 1998). Interestingly, the expression of *ruvA*, *ruvC* and *recG* in *M. tuberculosis* is reported to be increased in clinical lung samples (Rachman *et al.*, 2006b), and *recG* is also upregulated following infection of naïve and activated macrophages from wild-type and $NOS^{-/-}$ mice (Schnappinger *et al.*, 2003). The investigation of mutant strains of *M. tuberculosis* is awaited with interest.

7.7. RecN

The RecN protein plays an important role in the repair of double-stranded breaks, particularly if more than one break is present in the cell (Meddows *et al.*, 2005). An *E. coli recN* mutant strain exhibits a degree of susceptibility to agents causing double-strand breaks similar to that of a strain mutated in *recA* (Kosa *et al.*, 2004). The RecN protein is believed to have ATPase activity, but its biochemical properties have yet to be ascertained. The predicted secondary structure of RecN suggests that it may be a member of the structural maintenance of chromosomes protein family (Kidane *et al.*, 2004), a group of proteins involved in a number of aspects of DNA metabolism including chromosome condensation, sister chromatid cohesion and DNA repair (Hirano, 2002). Indeed, recent studies in *Bacillus subtilis* have revealed that RecN forms discrete foci following the induction of double-stranded breaks (Kidane *et al.*, 2004; Sanchez *et al.*, 2006). Homologues of RecN are found in *M. tuberculosis*, *M. bovis*, *M. marinum*, *M. avium* subsp. *paratuberculosis*, and *M. leprae*. In *M. tuberculosis* expression of the *recN* gene is increased in clinical lung samples of TB infected patients (Rachman *et al.*, 2006b), and the gene is predicted to be required for survival *in vivo* in a mouse model of infection (Sassetti and Rubin, 2003).

8. Non-Homologous End-Joining

Non-homologous end-joining (NHEJ) provides an alternative to homologous recombination for the repair of double-strand breaks, but as no homology is required to guide the process it can result in the addition or subtraction of nucleotides during end-processing and thus may be mutagenic. NHEJ is a major pathway for the repair of double-strand breaks in eukaryotes, but it is only relatively recently that this process has come to be recognised as a DNA repair pathway in prokaryotes as well. The breakthrough came in 2001 with the identification of protein homologues of a key component of this pathway known as Ku (Aravind and Koonin, 2001; Doherty *et al.*, 2001) swiftly followed by the demonstration of a functional NHEJ process in *Bacillus subtilis* and mycobacteria (Weller *et al.*, 2002). Unusually, it is in mycobacteria, along with *Pseudomonas*, that this repair pathway has been best characterised to date, perhaps because of its absence in *E. coli* K12.

The first step in this repair pathway is the binding of specific proteins to the DNA ends. This potentially serves two functions: to limit degradation and to act as an end-bridging factor to bring the two ends of a break together. Following this some kind of processing is commonly required to render the ends suitable for ligation. It is thought that the alignment of the ends is guided by regions of microhomology involving 1-4 nucleotides close to the break sites (Hefferin and Tomkinson, 2005). Resulting flaps are then removed or gaps filled in, prior to ligation to complete the repair (Figure 6).

In eukaryotes, NHEJ requires a number of proteins, including Ku70, Ku80, DNA-PKcs, DNA ligase IV, XRCC4 and Artemis (Doherty & Jackson, 2001). However, in prokaryotes it has been demonstrated that only two proteins, Ku and ligase, are required (Della *et al.*, 2004). Bacterial Ku is a much smaller protein that the eukaryotic Ku proteins, comprising only the central core domain required for dimerisation and DNA binding, and lacking other regions involved in protein interactions (Aravind and Koonin, 2001; Doherty *et al.*, 2001). In contrast, the mycobacterial ligase LigD is a multi-functional protein containing nuclease and polymerase domains in addition to the ligase domain. Intriguingly, the corresponding *Pseudomonas* ligase contains the same domains but in a different order, while in *Strepomyces coelicolor* each of these functions is provided by a separate protein (Bowater and Doherty, 2006).

Ku can bind not only to double-stranded DNA ends that are blunt but also to ends having a 3' or 5' overhang, and the *M. tuberculosis* Ku (Rv0937c) has been demonstrated to bind to such ends as a homodimer (Weller *et al.*, 2002). This study also showed that *in vitro* the number of Ku molecules bound is dependent upon the length of the DNA fragment in the assay, suggesting that following binding Ku can translocate along the DNA molecule, as has been found previously for eukaryotic Ku. These activities are understandable in the light of the crystal and predicted structures of Ku proteins from both domains of life (Bowater and Doherty, 2006; Walker *et al.*, 2001). The two subunits of Ku combine to form an asymmetric ring, which fits around the dsDNA molecule facilitating translocation. As the interactions between the protein and the DNA occur via the sugar-phosphate DNA backbone, binding is independent of DNA sequence.

In addition, *M. tuberculosis* Ku stimulated the ligation activity of its cognate ligase (Rv0938) but not that of unrelated ligases (Weller *et al.*, 2002). Although eukaryotic Ku likewise stimulates its corresponding ligase (ligase IV/XRCC4), these abilities were not interchangeable: mycobacterial Ku could not stimulate the activity of ligase IV/XRCC4 and eukaryotic Ku could not stimulate the activity of Rv0938. The explanation for this lies in the requirement for a direct physical interaction between the Ku and ligase proteins.

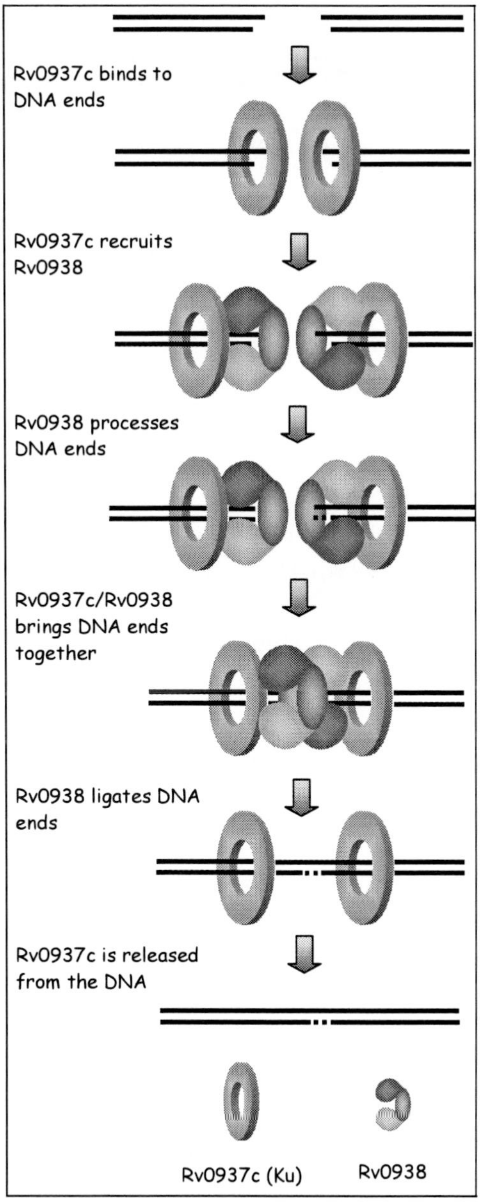

Figure 6. A schematic of the processes involved in non-homologous end-joining in *M. tuberculosis*. Double-stranded DNA ends are bound by dimers of the Ku homologue Rv0937c, protecting the DNA from degradation. Rv0938 is recruited and processes any short 5' or 3' overhangs via its polymerase or exonuclease domains respectively. The DNA ends are held in close proximity within the Rv0937c-Rv0938-DNA complex and ligated via the ligase domain of Rv0938. Finally, Rv0937c must be released from the repaired DNA, perhaps involving subunit dissociation.

Rv0938 (LigD) was originally identified as the ligase most probably involved in NHEJ owing to its genetic location adjacent to the gene Rv0937c encoding Ku (Weller *et al.*, 2002), with these two genes being divergently transcribed. However, *M. tuberculosis* also contains three other ligases: LigA encoded by Rv3014c which is a NAD^+-dependent ligase, and LigB (Rv3062) and LigC (Rv3731) which are ATP-dependent ligases as is LigD. The importance of LigD for NHEJ was confirmed by the analysis of *M. tuberculosis* strains deleted for each of *ligB*, *ligC* and *ligD* individually (*ligA* being essential) using an assay requiring the circularisation of a linearized plasmid (Gong *et al.*, 2004). Deletion of *ligB* or *ligC* had no effect on the efficiency of plasmid re-joining, whilst deletion of *ligD* resulted in a 70-79% decrease in the efficiency of NHEJ. This study also showed that there was no significant difference in the efficiency of NHEJ of DNA ends that were blunt or that had 5' overhangs.

The functionality of the nuclease and polymerase domains of Rv0938 was demonstrated in an elegant series of *in vitro* assays (Della *et al.*, 2004). Using a variety of oligonucleotide substrates designed to mimic microhomology-mediated joints typical of NHEJ, it was demonstrated that this single protein could fill in gaps and remove 3' single-stranded flaps without affecting the double-stranded regions of DNA, enabling the joining of incompatible ends. Furthermore, these activities were eliminated by mutation of conserved amino acids in the appropriate domains, confirming that the functions assayed were performed by Rv0938.

Homologues of both Ku and Rv0938 are present in *M. bovis*, *M. marinum* and *M. avium* subsp. *paratuberculosis* whilst the corresponding loci in *M. leprae* are pseudogenes. Intriguingly, whilst these divergently transcribed genes are adjacent in *M. tuberculosis*, they are separated by one and four unrelated open reading frames in *M. marinum* and *M. avium* subsp. *paratuberculosis* respectively (Figure 7). In contrast, the homologues of the gene Rv0939 that is located downstream of Rv0938 in *M. tuberculosis* remain adjacent to the ligase genes on the chromosomes of these species, raising the possibility that these two genes may have some kind of functional relationship. Functional homologues of Ku have also been identified in two mycobacteriophage termed Omega and Corndog, although these bacteriophage genomes lack DNA ligase genes (Pitcher *et al.*, 2006). Correspondingly, these mycobacteriophage depend on their host mycobacterial ligase for genome circularization and infectivity. It was proposed that the mycobacteriophage Ku may protect the phage DNA ends from degradation on entering the host, as well as functioning in ligation, as an explanation for the presence of Ku alone in these phages.

Studies of NHEJ in *M. smegmatis* have revealed that the process is highly mutagenic, having an error rate in the region of 50% (Gong *et al.*, 2005). In the case of blunt-ended substrates the majority (85%) of the imprecise joins contained a single nucleotide insertion, while when a substrate with a 5' overhang was used similar numbers of insertions, generated by fill-in DNA synthesis, and unidirectional deletions were formed. In the absence of Ku or LigD there was a marked decrease in the efficiency of repair, but a concomitant increase in the accuracy of the residual repair events. Elimination of just the polymerase activity of LigD by mutation increased the fidelity of repair of blunt-ended substrates as a consequence of preventing nucleotide insertions (Zhu and Shuman, 2006), while mutation of LigD that ablates only the ligase activity results in decreased fidelity (Akey *et al.*, 2006).

Both the efficiency and accuracy of repair were more dramatically affected in the strain deleted for Ku than in that lacking LigD, and although deletion of LigC had no detectable effect on its own, as seen for *M. tuberculosis*, deletion of LigC in a strain already lacking LigD caused a further decrease in repair efficiency but increase in accuracy with blunt-ended substrates, suggesting that LigC may play an auxiliary role in NHEJ. Intriguingly, the gene in *M. tuberculosis* located adjacent to LigC and divergently transcribed from it, Rv3730c, is similar to the amino-terminal primase domain of Rv0938, raising the possibility that it may function in conjunction with LigC in this process. This domain is present in only one other gene in *M. tuberculosis*, Rv0269c.

It has been suggested that NHEJ may provide an important alternative to homologous recombination for the repair of double-stranded breaks for mycobacteria in the dormant state (Gong *et al.*, 2005; Weller *et al.*, 2002). It is of interest in this regard that LigD is able to use ribonucleotides for gap synthesis as well as deoxyribonucleotides, as the dNTP pool might be limiting in such non-replicating cells (Della *et al.*, 2004; Gong *et al.*, 2005). However, this attractive possibility remains to be addressed experimentally.

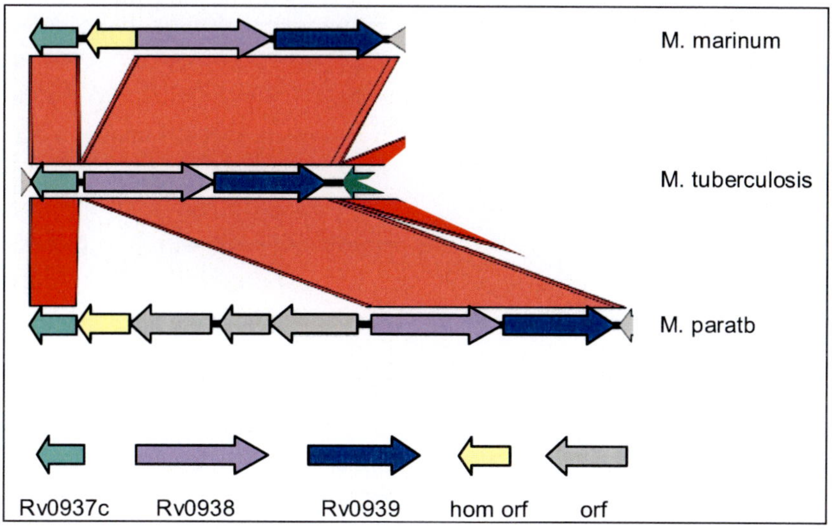

Figure 7. Arrangement of NHEJ loci in *M. tuberculosis*, *M. marinum* and *M. avium* subsp. *paratuberculosis*, based on PROmer alignments generated via the MycoDB website (http://myco.bham.ac.uk/). Whilst the Ku and ligase genes are adjacent in *M. tuberculosis*, in *M. marinum* they are separated by one additional open reading frame, and in *M. avium* subsp. *paratuberculosis* there are four orfs between the homologues of Rv0937c and Rv0938, one of which is highly similar to the intervening orf of *M. marinum*.

9. DNA-DAMAGE INDUCED MUTAGENESIS

Although the preferred response to DNA damage is to repair it, an alternative means of survival is provided by mechanisms to tolerate the damage. The majority of lesions will block replication by the replicative polymerase, PolIII, but most bacteria contain specialised polymerases which are able to copy the DNA template across the lesion, following which PolIII can resume synthesis. However, this translesion synthesis often results in the incorporation of incorrect bases resulting in mutation. For this reason the expression and sometimes the function of the enzymes involved are tightly regulated.

In *E. coli* there are three translesion polymerases: PolII, PolIV (Wagner *et al.*, 1999) and PolV (Tang *et al.*, 1999) encoded by *polB*, *dinB*, and *umuDC* respectively. The first of these is a member of the B family of polymerases, while the other two both belong to the Y family. The expression of all three of these polymerase is controlled by the SOS system (see below) and increases following DNA damage although with different kinetics, with PolV being subject to additional post-translational regulation (Gonzalez and Woodgate, 2002) that is beyond the scope of this review. These polymerases vary in their ability to direct synthesis past different types of damage and the particular lesions can influence whether this process is mutagenic or error-free (Napolitano *et al.*, 2000).

M. tuberculosis contains two identified members of the Y family of polymerases, termed *dinP* (Rv3056) and *dinX* (Rv1537). However, it has been demonstrated that DNA damage does not result in increased expression of either *dinP* or *dinX* (Boshoff *et al.*, 2003; Brooks *et al.*, 2001), making it unlikely that these enzymes play an important role in translesion synthesis. Rather it has been clearly shown that DNA damage inducible mutagenesis in *M. tuberculosis* and *M. smegmatis* is dependent on DnaE2 (Boshoff *et al.*, 2003), a member of the C family of polymerases normally represented by the replicative polymerase. The expression of *dnaE2* (Rv3370c) is highly induced following DNA damage under the control of the SOS system (Boshoff *et al.*, 2003; Davis *et al.*, 2002a; Rand *et al.*, 2003). Importantly, the mutation frequency increased following DNA damage in the wild-type mycobacterial strains but not in strains in which *dnaE2* had been inactivated, whilst in the complemented strains the frequency was restored to wild-type levels. The ability of *M. tuberculosis* to perform translesion synthesis mediated by DnaE2 was shown have a role in pathogenesis at late stages of infection in a mouse model, and, significantly, appeared to be important for the development of antibiotic resistance *in vivo* (Boshoff *et al.*, 2003). The involvement of DnaE in DNA damage inducible mutagenesis has since been demonstrated in other bacteria such as *Streptococcus pyogenes* (Bruck *et al.*, 2003), *Bacillus subtilis* (Le Chatelier *et al.*, 2004) and *Caulobacter cresentus* (Galhardo *et al.*, 2005). All three of these polymerases are conserved in *M. bovis* and *M. avium* subsp. *paratuberculosis*, whilst *dinP* is absent from *M. marinum* and all three genes are represented by pseudogenes in *M. leprae*. In *M. tuberculosis*, the expression of *dnaE2* was increased following infection of mice (Boshoff *et al.*, 2003) and all three polymerase genes have been reported to be expressed at an elevated level in clinical lung samples from TB patients (Rachman *et al.*, 2006b).

10. Other Potential DNA Repair Genes

As well as genes encoding proteins with a known role in DNA repair, there are a number of genes in both *E. coli* and *M. tuberculosis* for which the protein product is predicted to be involved in DNA metabolism but without a clear understanding of its function. In some cases this is based on the identification of conserved sequence motifs, while others are implicated in DNA repair as their expression is induced following DNA damage.

One such gene has been termed *lhr*, for long helicase related, as it was the longest known protein in *E. coli* (Reuven *et al.*, 1995). The protein encoded by this gene contains the seven conserved motifs of the DNA and RNA helicase superfamily II. Strains of *E. coli* in which *lhr* is mutated are not affected for growth, or sensitivity to UV irradiation or hydrogen peroxide (Reuven *et al.*, 1995). Despite additionally combining mutation of *lhr* with mutations in other genes coding for helicases, it was not possible to determine the physiological function of the Lhr protein. Homologues of *lhr* are present in *M. tuberculosis,*

M. bovis, *M. marinum* and *M. avium* subsp. *paratuberculosis* whilst the corresponding locus in *M. leprae* is a pseudogene. The expression of the *M. tuberculosis lhr* (Rv3296) is DNA damage inducible (Rand *et al.*, 2003) and has been reported be increased following exposure to hydrogen peroxide and intra-cellularly in mouse macrophages (Schnappinger *et al.*, 2003).

Another poorly characterized gene proposed to have a role in DNA repair in *E. coli* is *dinG*. This gene was originally identified as its promoter was DNA damage inducible (Lewis *et al.*, 1992), and again its sequence suggested that it encodes a DNA helicase belonging to superfamily II. Recent studies have confirmed that the DinG protein does in fact possess a 5'-3' DNA helicase activity (Voloshin *et al.*, 2003). However, its biological function remains unclear as neither deletion of *dinG* nor overexpression of the encoded protein produced a significant effect on growth or UV sensitivity. Homologues of *dinG* are identifiable in *M. tuberculosis, M. bovis, M. marinum* and *M. avium* subsp. *paratuberculosis* but not in *M. leprae*. However, the regulation of the *M. tuberculosis dinG* (Rv1329c) differs from its *E. coli* counterpart in that it is not DNA damage inducible (Brooks *et al.*, 2001). Despite this, expression of Rv1329c has been reported to be increased following infection of mouse macrophages (Rachman *et al.*, 2006a) and in clinical lung samples (Rachman *et al.*, 2006b).

Similarly the gene *dinF* was identified as DNA damage inducible in *E. coli* (Kenyon and Walker, 1980) but in this case its function is completely uncharacterized. A possible clue to its role may be derived from the observation that the encoded protein resembles a multidrug efflux pump domain. Homologues of *dinF* are found in *M. tuberculosis, M. bovis* and *M. marinum* but this gene appears to be absent from *M. avium* subsp. *paratuberculosis* and *M. leprae*. Although the *M. tuberculosis dinF* homologue, Rv2836c, was not identified as DNA damage inducible (Rand *et al.*, 2003), interestingly, its expression has been reported to be enhanced following infection of mouse macrophages (Rachman *et al.*, 2006a; Schnappinger *et al.*, 2003), SCID and Balb/c mice (Talaat *et al.*, 2004), and in clinical lung samples (Rachman *et al.*, 2006b), indicating that it may play an important role in infection.

Two proteins encoded by the *sbcC* and *sbcD* genes in *E. coli* form a complex that is responsible for preventing the replication of long palindromes via a double-stranded exonuclease activity. The SbcCD complex has been demonstrated to cleave hairpin DNA structures (Connelly *et al.*, 1998) such as would be formed by extrusion from palindromic sequences, the ensuing breaks then being substrates for recombinational repair (Cromie *et al.*, 2000). The SbcCD complex is structurally related to the structural maintenance of chromosomes family of proteins (Connelly *et al.*, 1998), and more recently has been found to additionally remove protein bound to DNA ends by incising the DNA (Connelly *et al.*, 2003). However, despite these seemingly important functions, there are no identifiable homologues of SbcCD in any of *M. tuberculosis, M. bovis, M. marinum, M. avium* subsp. *paratuberculosis* or *M. leprae*.

Some mycobacterial genes are thought to have a role in DNA repair based on the presence of conserved motifs. Thus, as discussed above, Rv2191, Rv3201c and Rv3202c may have roles in nucleotide excision repair as they contain UvrC- or UvrD-like domains, while it is possible that Rv0269c and Rv3730c might have some ancillary role in non-homologous end-joining. In addition, the protein encoded by Rv2090 contains a 5'-3' exonuclease domain, suggesting that it too is likely to be involved in DNA repair. Alternatively, some or all of these proteins may function in novel pathways of DNA repair in mycobacteria.

Whilst some of the most highly induced genes in *M. tuberculosis* following DNA damage have a known function in DNA repair, for example *recA* and *dnaE2*, others have no counterparts in *E. coli*, but this pattern of expression suggests that the proteins they encode may also play a role. Amongst this latter group are the genes Rv3395c and Rv3394c which form a two gene operon, and are conserved in *M. bovis*, *M. marinum* and *M. avium* subsp. *paratuberculosis* although pseudogenes are found in *M. leprae*. Although no recognisable domains are present in Rv3395c, Rv3394c possesses some similarity to the DinP domain of Y family polymerases, further supporting the possibility that these genes may function in DNA repair in some way.

Another mycobacterial specific gene showing DNA damage inducible expression at a high level in *M. tuberculosis* is Rv2719c, which contains a transmembrane segment and a LysM domain. This gene is conserved in *M. bovis*, *M. marinum*, *M. avium* subsp. *paratuberculosis* and *M. leprae*. It has recently been shown that overproduction of the protein encoded by Rv2719c in *M. tuberculosis* leads to filamentation of the bacterial cells, similar to that caused by exposure to DNA damaging agents (Chauhan *et al.*, 2006). Expression of Rv2719c was induced by oxidative and nitrosative stress as well as following infection of the macrophage-like cell line THP-1, and corresponding levels of filamentation were observed. It thus appears that the function of this protein is analogous to that of SulA in *E. coli* in that increased expression following DNA damage blocks cell division to allow time for DNA lesions to be repaired.

11. Regulation of Gene Expression Following DNA Damage

Although some DNA repair genes are constitutively expressed, a number are subject to regulation whereby the expression level is elevated following DNA damage. Such controlled expression is particularly important where the function of the gene product can be disadvantageous in the absence of damage, such as for error-prone DNA polymerases.

11.1. The SOS System

The best characterized mechanism for the regulated expression of DNA repair genes is the SOS system which was first described in *E. coli*. The key components controlling gene expression in this system are RecA and LexA. LexA is a repressor protein that in the undamaged state binds to a specific DNA motif, termed the SOS box, upstream of the genes it regulates (Walker, 1984). This motif is commonly found overlapping or between the promoter elements (Schnarr *et al.*, 1991) and LexA binding at this location is thought to reduce access by RNA polymerase, thus limiting transcription of the downstream gene. When the DNA becomes damaged regions of single-stranded DNA are formed, either as a consequence of replication blockage or processing by constitutive DNA repair components. The generation of single-stranded DNA provides a common signal for DNA damage of various kinds and is detected by RecA. RecA binds to single-stranded DNA to form a nucleoprotein filament, and this 'activated' form stimulates the auto-catalytic cleavage of LexA (Little, 1991). The cleaved fragments of LexA no longer bind with significant affinity to the SOS box, resulting in the alleviation of repression and so increased expression of the SOS regulon. Not all genes in the regulon will be induced to the same extent as the degree of induction for an individual gene will depend on a number of factors such as the affinity of LexA for the particular SOS box, the location of the SOS box relative to the promoter, and the promoter strength.

The basic principles of this regulatory mechanism are found in many other species of bacteria, although the DNA sequence of the LexA binding site, or SOS box, varies. Thus, whilst the SOS box in *E. coli* and other enterobacteria has the consensus sequence taCTGTatatatatACAGta (Lewis *et al.*, 1994), in the best characterized Gram-positive example, *Bacillus subtilis*, the sequence CGAACRNRYGTTCG (Winterling *et al.*, 1998) is recognised by the LexA homologue known as DinR. The SOS system of gene regulation has been demonstrated to be conserved and functional in mycobacteria. A LexA homologue was identified in *M. tuberculosis*, *M. leprae* and *M. smegmatis* (Durbach *et al.*, 1997; Movahedzadeh *et al.*, 1997b) and LexA is also present in *M. bovis*, *M. marinum* and *M. avium* subsp. *paratuberculosis*. Following expression and purification both the *M. leprae* and *M. tuberculosis* proteins bound to a specific sequence with similarity to the *B. subtilis* SOS box upstream of *recA* (Durbach *et al.*, 1997; Movahedzadeh *et al.*, 1997a). Subsequently, the consensus sequence for the mycobacterial SOS box was defined by a systematic mutagenesis approach as TCGAACnnnnGTTCGA (Davis *et al.*, 2002a).

This, coupled with expression data, allowed the identification of the *M. tuberculosis* SOS regulon (Davis *et al.*, 2002a). Surprisingly, a much smaller number of genes are regulated by LexA in *M. tuberculosis* than in *E. coli.* A total of sixteen genes appear to be regulated by this classical mechanism of DNA damage induction, and of these only two have a known role in DNA repair apart from *recA* and *lexA* themselves. Interestingly, one of these is *dnaE2* which has been shown to be the key mediator of DNA damage inducible mutagenesis in *M. tuberculosis*, while the other is *ruvC*, the Holliday junction resolvase important in recombination. In contrast, over 40 genes are regulated by the SOS system in *E. coli* (Courcelle *et al.*, 2001; Fernandez de Henestrosa *et al.*, 2000) of which approximately one-third are known to have a role in DNA repair.

11.2. Alternative Mechanism(s) of Regulation

Although few DNA repair genes were found to be part of the SOS regulon in *M. tuberculosis*, analysis of a selection of genes with diverse roles in DNA metabolism that are known to be DNA damage inducible in *E. coli* revealed that a number of them were also induced following DNA damage in *M. tuberculosis* (Brooks *et al.*, 2001). However, in two cases binding of LexA to the appropriate upstream region could not be detected, providing the first indication that an alternative mode of regulation might exist.

Table 2. Induction of expression of DNA repair genes in *M. tuberculosis* following DNA damage or infection

Gene name	Rv No.	Mitomycin C[1]	UV irradiation[2]	H_2O_2[3]	Infection of macrophages[4]	Infection of humans[5]
Direct reversal						
ogt	Rv1316c	+	+	+		
NER						
uvrA	Rv1638	+	+	+		+
uvrB	Rv1633	+	+	+		
uvrC	Rv1420					+
uvrD1	Rv0949	+				+
uvrD2	Rv3198c	+	+			
mfd	Rv1020				+	
ercc3	Rv0861c	+				
BER						
alkA	Rv1317c	+	+	+	+	
end	Rv0670					
fpg	Rv2924c					
	Rv0944					
	Rv2464c					+
mpg	Rv1688					+
mutY	Rv3589					
nei	Rv3297	+				
nth	Rv3674c					
tagA	Rv1210	+				
ung	Rv2976c	+		+		
xthA	Rv0427c	+				
Recombination						
recA	Rv2737c	+	+	+		
recB	Rv0630c					+
recC	Rv0631c				+	
recD	Rv0629c					
recF	Rv0003					
recG	Rv2973c				+	+
recN	Rv1696					+
recO	Rv2362c	+				
recR	Rv3715c					
recX	Rv2736c	+	+	+		
ruvA	Rv2593c	+	+			+
ruvB	Rv2592c	+				
ruvC	Rv2594c	+	+			+

Table 2. Continued

Gene name	Rv No.	Mitomycin C[1]	UV irradiation[2]	H_2O_2[3]	Infection of macrophages[4]	Infection of humans[5]
NHEJ						
Ku	Rv0937c					+
ligD	Rv0938					
Mutagenesis						
dinP	Rv3056					+
dinX	Rv1537					+
dnaE2	Rv3370c	+	+			+
Other potential DNA repair genes						
dinF	Rv2836c				+	+
dinG	Rv1329c				+	+
lhr	Rv3296	+	+		+	
ligB	Rv3062	+				
ligC	Rv3731					
radA	Rv3585	+	+	+	+	+
xseA	Rv1108c					
xseB	Rv1107c					
	Rv0269c					
	Rv2090					+
	Rv2191	+			+	
	Rv2719c	+	+	+		
	Rv3201c	+				+
	Rv3202c	+	+	+	+	
	Rv3394c	+	+	+		
	Rv3395c	+	+			
	Rv3730c					
General						
ligA	Rv3014c					
polA	Rv1629	+				
ssb	Rv0054	+		+		+
Regulation						
lexA	Rv2720	+	+			

[1]Combined data from (Boshoff *et al.*, 2003) and (Rand *et al.*, 2003).

[2]Data from (Boshoff *et al.*, 2003).

[3]Combined data from (Boshoff *et al.*, 2003) and (Schnappinger *et al.*, 2003).

[4]Combined data from (Schnappinger *et al.*, 2003), (Cappelli *et al.*, 2006) and (Rachman *et al.*, 2006a).

[5]Data from (Rachman *et al.*, 2006b).

Definitive proof that another mechanism for DNA damage induction of gene expression must exist in mycobacteria was provided by a comparative analysis of the two promoters for the *M. tuberculosis recA* gene (Davis *et al.*, 2002b). One promoter (P2) is regulated by LexA and as expected exhibits high-level constitutive expression when a mutation is introduced into the SOS box that prevents LexA binding. Conversely, expression of a non-cleavable form of LexA prevents induction, and in the absence of RecA there is no induction as regulation by LexA is dependent on RecA. In contrast, the other promoter (P1) is DNA damage inducible under all of these conditions.

Comparison of the expression profile following DNA damage in wild-type and *recA* defective *M. tuberculosis* strains revealed that the majority of DNA damage inducible genes remained inducible in the absence of RecA (Rand *et al.*, 2003). Notably, this included genes involved in a number of DNA repair pathways. The alternative mechanism or mechanisms involved have yet to be elucidated, but interestingly a DNA sequence motif common to a number of these genes has been identified in the regions upstream of their coding sequences with features that suggest that it may represent a promoter (Gamulin *et al.*, 2004). Thus, one possibility is that the expression of at least a subset of the RecA-independent DNA-damage inducible genes is controlled by a specific sigma factor.

12. Conclusion

In conclusion, many aspects of DNA repair in mycobacteria resemble those described in other species of bacteria. The core processes of nucleotide excision repair, base excision repair and recombination are all present, as is non-homologous end-joining, a repair pathway absent from *E. coli* (Table 1). Nevertheless, there are some notable differences such as the absence of a mismatch repair system and the requirement of a different class of DNA polymerase for inducible mutagenesis. Even within a conserved repair pathway there appear to be subtle variations on established pathways, for example the proposed function of Rv2191. Intriguingly, mycobacteria appear to possess multiple genes that resemble certain DNA repair genes such as *uvrD* and *fpg* suggesting that DNA repair may be particularly important for mycobacterial survival, at least under some conditions. Furthermore, the increase in expression observed for a number of DNA repair genes in infection models (Table 2) raises the possibility of an involvement in pathogenesis.

Acknowledgements

This work was supported by the Medical Research Council UK.

References

Akey, D., Martins, A., Aniukwu, J., Glickman, M.S., Shuman, S., and Berger, J.M. (2006) Crystal structure and nonhomologous end-joining function of the ligase component of Mycobacterium DNA ligase D. *J Biol Chem 281*: 13412-13423.

Allen, D.J., Makhov, A., Grilley, M., Taylor, J., Thresher, R., Modrich, P., and Griffith, J.D. (1997) MutS mediates heteroduplex loop formation by a translocation mechanism. *EMBO J 16*: 4467-4476.

Anderson, D.G., and Kowalczykowski, S.C. (1997a) The recombination hot spot chi is a regulatory element that switches the polarity of DNA degradation by the RecBCD enzyme. *Genes Dev 11*: 571-581.

Anderson, D.G., and Kowalczykowski, S.C. (1997b) The translocating RecBCD enzyme stimulates recombination by directing RecA protein onto ssDNA in a chi-regulated manner. *Cell 90*: 77-86.

Aravind, L., and Koonin, E.V. (2001) Prokaryotic homologs of the eukaryotic DNA-end-binding protein Ku, novel domains in the Ku protein and prediction of a prokaryotic double-strand break repair system. *Genome Res 11*: 1365-1374.

Asad, N.R., de Almeida, C.E., Asad, L.M., Felzenszwalb, I., and Leitao, A.C. (1995) Fpg and UvrA proteins participate in the repair of DNA lesions induced by hydrogen peroxide in low iron level in *Escherichia coli*. *Biochimie 77*: 262-264.

Beam, C.E., Saveson, C.J., and Lovett, S.T. (2002) Role for *radA/sms* in Recombination Intermediate Processing in *Escherichia coli*. *J. Bacteriol. 184*: 6836-6844.

Blaisdell, J.O., Hatahet, Z., and Wallace, S.S. (1999) A novel role for *Escherichia coli* endonuclease VIII in prevention of spontaneous G-->T transversions. *J Bacteriol 181*: 6396-6402.

Bootsma, D., Weeda, G., Vermeulen, W., van Vuuren, H., Troelstra, C., van der Spek, P., and Hoeijmakers, J. (1995) Nucleotide excision repair syndromes: molecular basis and clinical symptoms. *Philos Trans R Soc Lond B Biol Sci 347*: 75-81.

Bork, J.M., Cox, M.M., and Inman, R.B. (2001) The RecOR proteins modulate RecA protein function at 5' ends of single-stranded DNA. *EMBO J 20*: 7313-7322.

Boshoff, H.I., Reed, M.B., Barry, C.E., 3rd, and Mizrahi, V. (2003) DnaE2 polymerase contributes to in vivo survival and the emergence of drug resistance in *Mycobacterium tuberculosis*. *Cell 113*: 183-193.

Bowater, R., and Doherty, A.J. (2006) Making ends meet: repairing breaks in bacterial DNA by non-homologous end-joining. *PLoS Genet 2*: e8.

Brooks, P.C., Movahedzadeh, F., and Davis, E.O. (2001) Identification of some DNA damage-inducible genes of *Mycobacterium tuberculosis*: apparent lack of correlation with LexA binding. *J Bacteriol 183*: 4459-4467.

Bruck, I., Goodman, M.F., and O'Donnell, M. (2003) The essential C family DnaE polymerase is error-prone and efficient at lesion bypass. *J Biol Chem 278*: 44361-44368.

Buchmeier, N.A., Lipps, C.J., So, M.Y., and Heffron, F. (1993) Recombination-deficient mutants of *Salmonella typhimurium* are avirulent and sensitive to the oxidative burst of macrophages. *Mol Microbiol 7*: 933-936.

Cadman, C.J., Matson, S.W., and McGlynn, P. (2006) Unwinding of Forked DNA Structures by UvrD. *J Mol Biol 362*: 18-25.

Cappelli, G., Volpe, E., Grassi, M., Liseo, B., Colizzi, V., and Mariani, F. (2006) Profiling of *Mycobacterium tuberculosis* gene expression during human macrophage infection: upregulation of the alternative sigma factor G, a group of transcriptional regulators, and proteins with unknown function. *Res Microbiol 157*: 445-455.

Chauhan, A., Lofton, H., Maloney, E., Moore, J., Fol, M., Madiraju, M.V., and Rajagopalan, M. (2006) Interference of *Mycobacterium tuberculosis* cell division by Rv2719c, a cell wall hydrolase. *Mol Microbiol 62*: 132-147.

Chedin, F., and Kowalczykowski, S.C. (2002) A novel family of regulated helicases/nucleases from Gram-positive bacteria: insights into the initiation of DNA recombination. *Mol Microbiol 43*: 823-834.

Chow, K.H., and Courcelle, J. (2004) RecO acts with RecF and RecR to protect and maintain replication forks blocked by UV-induced DNA damage in *Escherichia coli*. *J Biol Chem 279*: 3492-3496.

Clark, A.J. (1973) Recombination deficient mutants of *E. coli* and other bacteria. *Annu Rev Genet 7*: 67-86.

Claverys, J.P., and Lacks, S.A. (1986) Heteroduplex deoxyribonucleic acid base mismatch repair in bacteria. *Microbiol Rev 50*: 133-165.

Cole, S.T., Brosch, R., Parkhill, J., Garnier, T., Churcher, C., Harris, D., Gordon, S.V., Eiglmeier, K., Gas, S., Barry, C.E., 3rd, Tekaia, F., Badcock, K., Basham, D., Brown, D., Chillingworth, T., Connor, R., Davies, R., Devlin, K., Feltwell, T., Gentles, S., Hamlin, N., Holroyd, S., Hornsby, T., Jagels, K., Barrell, B.G., and et al. (1998) Deciphering the biology of *Mycobacterium tuberculosis* from the complete genome sequence. *Nature 393*: 537-544.

Connelly, J.C., Kirkham, L.A., and Leach, D.R. (1998) The SbcCD nuclease of *Escherichia coli* is a structural maintenance of chromosomes (SMC) family protein that cleaves hairpin DNA. *Proc Natl Acad Sci U S A 95*: 7969-7974.

Connelly, J.C., de Leau, E.S., and Leach, D.R. (2003) Nucleolytic processing of a protein-bound DNA end by the *E. coli* SbcCD (MR) complex. *DNA Repair (Amst) 2*: 795-807.

Courcelle, J., Khodursky, A., Peter, B., Brown, P.O., and Hanawalt, P.C. (2001) Comparative gene expression profiles following UV exposure in wild-type and SOS-deficient *Escherichia coli*. *Genetics 158*: 41-64.

Courcelle, J., and Hanawalt, P.C. (2003) RecA-dependent recovery of arrested DNA replication forks. *Annu Rev Genet 37*: 611-646.

Cox, M.M. (1991) The RecA protein as a recombinational repair system. *Mol Microbiol 5*: 1295-1299.

Cromie, G.A., Millar, C.B., Schmidt, K.H., and Leach, D.R. (2000) Palindromes as substrates for multiple pathways of recombination in *Escherichia coli*. *Genetics 154*: 513-522.

Curti, E., Smerdon, S.J., and Davis, E.O. (2007) Characterisation of the helicase activity and substrate specificity of *Mycobacterium tuberculosis* UvrD. *J. Bacteriol. 189: 1542-1555.*

Dao, V., and Modrich, P. (1998) Mismatch-, MutS-, MutL-, and helicase II-dependent unwinding from the single-strand break of an incised heteroduplex. *J Biol Chem 273*: 9202-9207.

Darwin, K.H., and Nathan, C.F. (2005) Role for nucleotide excision repair in virulence of *Mycobacterium tuberculosis*. *Infect Immun 73*: 4581-4587.

Datta, S., Prabu, M.M., Vaze, M.B., Ganesh, N., Chandra, N.R., Muniyappa, K., and Vijayan, M. (2000) Crystal structures of *Mycobacterium tuberculosis* RecA and its complex with ADP-AlF(4): implications for decreased ATPase activity and molecular aggregation. *Nucleic Acids Res 28*: 4964-4973.

Datta, S., Ganesh, N., Chandra, N.R., Muniyappa, K., and Vijayan, M. (2003) Structural studies on MtRecA-nucleotide complexes: insights into DNA and nucleotide binding and the structural signature of NTP recognition. *Proteins 50*: 474-485.

Davis, E.O., Sedgwick, S.G., and Colston, M.J. (1991) Novel structure of the *recA* locus of *Mycobacterium tuberculosis* implies processing of the gene product. *J Bacteriol 173*: 5653-5662.

Davis, E.O., Jenner, P.J., Brooks, P.C., Colston, M.J., and Sedgwick, S.G. (1992) Protein splicing in the maturation of *M. tuberculosis* RecA protein: a mechanism for tolerating a novel class of intervening sequence. *Cell 71*: 201-210.

Davis, E.O., Dullaghan, E.M., and Rand, L. (2002a) Definition of the Mycobacterial SOS Box and Use To Identify LexA-Regulated Genes in *Mycobacterium tuberculosis. J. Bacteriol. 184*: 3287-3295.

Davis, E.O., Springer, B., Gopaul, K.K., Papavinasasundaram, K.G., Sander, P., and Bottger, E.C. (2002b) DNA damage induction of *recA* in *Mycobacterium tuberculosis* independently of RecA and LexA. *Mol Microbiol 46*: 791-800.

De Mot, R., Schoofs, G., and Vanderleyden, J. (1994) A putative regulatory gene downstream of *recA* is conserved in gram-negative and gram-positive bacteria. *Nucleic Acids Res 22*: 1313-1314.

Della, M., Palmbos, P.L., Tseng, H.M., Tonkin, L.M., Daley, J.M., Topper, L.M., Pitcher, R.S., Tomkinson, A.E., Wilson, T.E., and Doherty, A.J. (2004) Mycobacterial Ku and ligase proteins constitute a two-component NHEJ repair machine. *Science 306*: 683-685.

Demple, B., and Harrison, L. (1994) Repair of oxidative damage to DNA: enzymology and biology. *Annu Rev Biochem 63*: 915-948.

Dillingham, M.S., Spies, M., and Kowalczykowski, S.C. (2003) RecBCD enzyme is a bipolar DNA helicase. *Nature 423*: 893-897.

Dixon, D.A., and Kowalczykowski, S.C. (1993) The recombination hotspot chi is a regulatory sequence that acts by attenuating the nuclease activity of the *E. coli* RecBCD enzyme. *Cell 73*: 87-96.

Dizdaroglu, M., Bauche, C., Rodriguez, H., and Laval, J. (2000) Novel substrates of *Escherichia coli* Nth protein and its kinetics for excision of modified bases from DNA damaged by free radicals. *Biochemistry 39*: 5586-5592.

Doherty, A.J., Jackson, S.P., and Weller, G.R. (2001) Identification of bacterial homologues of the Ku DNA repair proteins. *FEBS Lett 500*: 186-188.

Drees, J.C., Lusetti, S.L., Chitteni-Pattu, S., Inman, R.B., and Cox, M.M. (2004) A RecA filament capping mechanism for RecX protein. *Mol Cell 15*: 789-798.

Dunderdale, H.J., Benson, F.E., Parsons, C.A., Sharples, G.J., Lloyd, R.G., and West, S.C. (1991) Formation and resolution of recombination intermediates by *E. coli* RecA and RuvC proteins. *Nature 354*: 506-510.

Durbach, S.I., Andersen, S.J., and Mizrahi, V. (1997) SOS induction in mycobacteria: analysis of the DNA-binding activity of a LexA-like repressor and its role in DNA damage induction of the *recA* gene from *Mycobacterium smegmatis*. *Mol Microbiol 26*: 643-653.

Durbach, S.I., Springer, B., Machowski, E.E., North, R.J., Papavinasasundaram, K.G., Colston, M.J., Bottger, E.C., and Mizrahi, V. (2003) DNA alkylation damage as a sensor of nitrosative stress in *Mycobacterium tuberculosis*. *Infect Immun 71*: 997-1000.

Fernandez de Henestrosa, A.R., Ogi, T., Aoyagi, S., Chafin, D., Hayes, J.J., Ohmori, H., and Woodgate, R. (2000) Identification of additional genes belonging to the LexA regulon in *Escherichia coli*. *Mol Microbiol 35*: 1560-1572.

Fishel, R. (1998) Mismatch repair, molecular switches, and signal transduction. *Genes Dev 12*: 2096-2101.

Friedberg, E., Walker, G., and Siede, W. (1995) *DNA repair and mutagenesis*. Washington, DC: American Society for Microbiology Press.

Frischkorn, K., Sander, P., Scholz, M., Teschner, K., Prammananan, T., and Bottger, E.C. (1998) Investigation of mycobacterial *recA* function: protein introns in the RecA of pathogenic mycobacteria do not affect competency for homologous recombination. *Mol Microbiol 29*: 1203-1214.

Galhardo, R.S., Almeida, C.E., Leitao, A.C., and Cabral-Neto, J.B. (2000) Repair of DNA lesions induced by hydrogen peroxide in the presence of iron chelators in *Escherichia coli*: participation of endonuclease IV and Fpg. *J Bacteriol 182*: 1964-1968.

Galhardo, R.S., Rocha, R.P., Marques, M.V., and Menck, C.F. (2005) An SOS-regulated operon involved in damage-inducible mutagenesis in *Caulobacter crescentus*. *Nucleic Acids Res 33*: 2603-2614.

Gamulin, V., Cetkovic, H., and Ahel, I. (2004) Identification of a promoter motif regulating the major DNA damage response mechanism of *Mycobacterium tuberculosis*. *FEMS Microbiol Lett 238*: 57-63.

Gong, C., Martins, A., Bongiorno, P., Glickman, M., and Shuman, S. (2004) Biochemical and genetic analysis of the four DNA ligases of mycobacteria. *J Biol Chem 279*: 20594-20606.

Gong, C., Bongiorno, P., Martins, A., Stephanou, N.C., Zhu, H., Shuman, S., and Glickman, M.S. (2005) Mechanism of nonhomologous end-joining in mycobacteria: a low-fidelity repair system driven by Ku, ligase D and ligase C. *Nat Struct Mol Biol 12*: 304-312.

Gonzalez, M., and Woodgate, R. (2002) The "tale" of UmuD and its role in SOS mutagenesis. *Bioessays 24*: 141-148.

Graham, J.E., and Clark-Curtiss, J.E. (1999) Identification of *Mycobacterium tuberculosis* RNAs synthesized in response to phagocytosis by human macrophages by selective capture of transcribed sequences (SCOTS). *Proc Natl Acad Sci U S A 96*: 11554-11559.

Gros, L., Saparbaev, M.K., and Laval, J. (2002) Enzymology of the repair of free radicals-induced DNA damage. *Oncogene 21*: 8905-8925.

Hefferin, M.L., and Tomkinson, A.E. (2005) Mechanism of DNA double-strand break repair by non-homologous end joining. *DNA Repair (Amst) 4*: 639-648.

Hirano, T. (2002) The ABCs of SMC proteins: two-armed ATPases for chromosome condensation, cohesion, and repair. *Genes Dev 16*: 399-414.

Howard-Flanders, P., Boyce, R.P., and Theriot, L. (1966) Three loci in *Escherichia coli* K-12 that control the excision of pyrimidine dimers and certain other mutagen products from DNA. *Genetics 53*: 1119-1136.

Ischenko, A.A., and Saparbaev, M.K. (2002) Alternative nucleotide incision repair pathway for oxidative DNA damage. *Nature 415*: 183-187.

Ishchenko, A.A., Deprez, E., Maksimenko, A., Brochon, J.C., Tauc, P., and Saparbaev, M.K. (2006) Uncoupling of the base excision and nucleotide incision repair pathways reveals their respective biological roles. *Proc Natl Acad Sci U S A 103*: 2564-2569.

Iyer, R.R., Pluciennik, A., Burdett, V., and Modrich, P.L. (2006) DNA mismatch repair: functions and mechanisms. *Chem Rev 106*: 302-323.

Jiang, D., Hatahet, Z., Blaisdell, J.O., Melamede, R.J., and Wallace, S.S. (1997a) *Escherichia coli* endonuclease VIII: cloning, sequencing, and overexpression of the *nei* structural gene and characterization of *nei* and *nei nth* mutants. *J Bacteriol 179*: 3773-3782.

Jiang, D., Hatahet, Z., Melamede, R.J., Kow, Y.W., and Wallace, S.S. (1997b) Characterization of *Escherichia coli* endonuclease VIII. *J Biol Chem 272*: 32230-32239.

Jurado, J., Maciejewska, A., Krwawicz, J., Laval, J., and Saparbaev, M.K. (2004) Role of mismatch-specific uracil-DNA glycosylase in repair of 3,N4-ethenocytosine in vivo. *DNA Repair (Amst) 3*: 1579-1590.

Kenyon, C.J., and Walker, G.C. (1980) DNA-damaging agents stimulate gene expression at specific loci in *Escherichia coli*. *Proc Natl Acad Sci U S A 77*: 2819-2823.

Kidane, D., Sanchez, H., Alonso, J.C., and Graumann, P.L. (2004) Visualization of DNA double-strand break repair in live bacteria reveals dynamic recruitment of *Bacillus subtilis* RecF, RecO and RecN proteins to distinct sites on the nucleoids. *Mol Microbiol 52*: 1627-1639.

Kosa, J.L., Zdraveski, Z.Z., Currier, S., Marinus, M.G., and Essigmann, J.M. (2004) RecN and RecG are required for *Escherichia coli* survival of Bleomycin-induced damage. *Mutat Res 554*: 149-157.

Kow, Y.W. (2002) Repair of deaminated bases in DNA. *Free Radic Biol Med 33*: 886-893.

Kowalczykowski, S., Dixon, D., Eggleston, A., Lauder, S., and Rehrauer, W. (1994) Biochemistry of homologous recombination in *Escherichia coli*. *Microbiol. Rev. 58*: 401-465.

Kowalczykowski, S.C. (2000) Initiation of genetic recombination and recombination-dependent replication. *Trends Biochem Sci 25*: 156-165.

Krokan, H.E., Standal, R., and Slupphaug, G. (1997) DNA glycosylases in the base excision repair of DNA. *Biochem J 325 (Pt 1)*: 1-16.

Kuemmerle, N.B., and Masker, W.E. (1980) Effect of the *uvrD* mutation on excision repair. *J Bacteriol 142*: 535-546.

Kumar, R.A., Vaze, M.B., Chandra, N.R., Vijayan, M., and Muniyappa, K. (1996) Functional characterization of the precursor and spliced forms of RecA protein of *Mycobacterium tuberculosis*. *Biochemistry 35*: 1793-1802.

Langle-Rouault, F., Maenhaut-Michel, G., and Radman, M. (1987) GATC sequences, DNA nicks and the MutH function in *Escherichia coli* mismatch repair. *EMBO J 6*: 1121-1127.

Le Chatelier, E., Becherel, O.J., d'Alencon, E., Canceill, D., Ehrlich, S.D., Fuchs, R.P., and Janniere, L. (2004) Involvement of DnaE, the second replicative DNA polymerase from *Bacillus subtilis*, in DNA mutagenesis. *J Biol Chem 279*: 1757-1767.

Lewis, L.K., Jenkins, M.E., and Mount, D.W. (1992) Isolation of DNA damage-inducible promoters in *Escherichia coli*: regulation of *polB* (*dinA*), *dinG*, and *dinH* by LexA repressor. *J Bacteriol 174*: 3377-3385.

Lewis, L.K., Harlow, G.R., Gregg-Jolly, L.A., and Mount, D.W. (1994) Identification of high affinity binding sites for LexA which define new DNA damage-inducible genes in *Escherichia coli*. *J Mol Biol 241*: 507-523.

Little, J.W. (1991) Mechanism of specific LexA cleavage: autodigestion and the role of RecA coprotease. *Biochimie 73*: 411-421.

Ljungquist, S., Lindahl, T., and Howard-Flanders, P. (1976) Methyl methane sulfonate-sensitive mutant of *Escherichia coli* deficient in an endonuclease specific for apurinic sites in deoxyribonucleic acid. *J Bacteriol 126*: 646-653.

Lloyd, R.G., Low, B., Godson, G.N., and Birge, E.A. (1974) Isolation and characterization of an *Escherichia coli* K-12 mutant with a temperature-sensitive *recA*- phenotype. *J Bacteriol 120*: 407-415.

Lloyd, R.G. (1991) Conjugational recombination in resolvase-deficient *ruvC* mutants of *Escherichia coli* K-12 depends on *recG*. *J Bacteriol 173*: 5414-5418.

Lloyd, R.G., and Sharples, G.J. (1993) Dissociation of synthetic Holliday junctions by *E. coli* RecG protein. *EMBO J 12*: 17-22.

Lobner-Olesen, A., Skovgaard, O., and Marinus, M.G. (2005) Dam methylation: coordinating cellular processes. *Curr Opin Microbiol 8*: 154-160.

Lusetti, S.L., Hobbs, M.D., Stohl, E.A., Chitteni-Pattu, S., Inman, R.B., Seifert, H.S., and Cox, M.M. (2006) The RecF protein antagonizes RecX function via direct interaction. *Mol Cell. 21*: 41-50.

McGlynn, P., and Lloyd, R.G. (2002) Genome stability and the processing of damaged replication forks by RecG. *Trends Genet 18*: 413-419.

Mechanic, L.E., Frankel, B.A., and Matson, S.W. (2000) *Escherichia coli* MutL loads DNA helicase II onto DNA. *J Biol Chem 275*: 38337-38346.

Meddows, T.R., Savory, A.P., Grove, J.I., Moore, T., and Lloyd, R.G. (2005) RecN protein and transcription factor DksA combine to promote faithful recombinational repair of DNA double-strand breaks. *Mol Microbiol 57*: 97-110.

Melamede, R.J., Hatahet, Z., Kow, Y.W., Ide, H., and Wallace, S.S. (1994) Isolation and characterization of endonuclease VIII from *Escherichia coli*. *Biochemistry 33*: 1255-1264.

Mendonca, V., Klepin, H., and Matson, S. (1995) DNA helicases in recombination and repair: construction of a delta *uvrD* delta *helD* delta *recQ* mutant deficient in recombination and repair. *J. Bacteriol. 177*: 1326-1335.

Mendonca, V.M., Kaiser-Rogers, K., and Matson, S.W. (1993) Double helicase II (*uvrD*)-helicase IV (*helD*) deletion mutants are defective in the recombination pathways of *Escherichia coli*. *J Bacteriol 175*: 4641-4651.

Mendonca, V.M., and Matson, S.W. (1995) Genetic analysis of delta *helD* and delta *uvrD* mutations in combination with other genes in the RecF recombination pathway in *Escherichia coli*: suppression of a *ruvB* mutation by a *uvrD* deletion. *Genetics 141*: 443-452.

Mizrahi, V., and Andersen, S. (1998) DNA repair in *Mycobacterium tuberculosis*. What have we learnt from the genome sequence? *Molecular Microbiology 29*: 1331-1339.

Mokkapati, S.K., Fernandez de Henestrosa, A.R., and Bhagwat, A.S. (2001) *Escherichia coli* DNA glycosylase Mug: a growth-regulated enzyme required for mutation avoidance in stationary-phase cells. *Mol Microbiol 41*: 1101-1111.

Moolenaar, G.F., Moorman, C., and Goosen, N. (2000) Role of the *Escherichia coli* nucleotide excision repair proteins in DNA replication. *J Bacteriol 182*: 5706-5714.

Moolenaar, G.F., van Rossum-Fikkert, S., van Kesteren, M., and Goosen, N. (2002) Cho, a second endonuclease involved in *Escherichia coli* nucleotide excision repair. *Proc Natl Acad Sci U S A 99*: 1467-1472.

Morimatsu, K., and Kowalczykowski, S.C. (2003) RecFOR proteins load RecA protein onto gapped DNA to accelerate DNA strand exchange: a universal step of recombinational repair. *Mol Cell 11*: 1337-1347.

Movahedzadeh, F., Colston, M., and Davis, E. (1997a) Determination of DNA sequences required for regulated *Mycobacterium tuberculosis* RecA expression in response to DNA-damaging agents suggests that two modes of regulation exist. *J. Bacteriol. 179*: 3509-3518.

Movahedzadeh, F., Colston, M.J., and Davis, E.O. (1997b) Characterization of *Mycobacterium tuberculosis* LexA: recognition of a Cheo (*Bacillus*-type SOS) box. *Microbiology 143*: 929-936.

Muniyappa, K., Vaze, M.B., Ganesh, N., Sreedhar Reddy, M., Guhan, N., and Venkatesh, R. (2000) Comparative genomics of *Mycobacterium tuberculosis* and *Escherichia coli* for recombination (*rec*) genes. *Microbiology 146*: 2093-2095.

Napolitano, R., Janel-Bintz, R., Wagner, J., and Fuchs, R.P. (2000) All three SOS-inducible DNA polymerases (Pol II, Pol IV and Pol V) are involved in induced mutagenesis. *EMBO J 19*: 6259-6265.

Ogawa, H. (1970) Genetic locations of *uvrD* and *pol* genes of *E. coli*. *Mol Gen Genet 108*: 378-381.

Pages, V., Koffel-Schwartz, N., and Fuchs, R.P. (2003) *recX*, a new SOS gene that is co-transcribed with the *recA* gene in *Escherichia coli*. *DNA Repair (Amst) 2*: 273-284.

Papavinasasundaram, K.G., Movahedzadeh, F., Keer, J.T., Stoker, N.G., Colston, M.J., and Davis, E.O. (1997) Mycobacterial *recA* is cotranscribed with a potential regulatory gene called *recX*. *Mol Microbiol 24*: 141-153.

Papavinasasundaram, K.G., Colston, M.J., and Davis, E.O. (1998) Construction and complementation of a *recA* deletion mutant of *Mycobacterium smegmatis* reveals that the intein in *Mycobacterium tuberculosis recA* does not affect RecA function. *Mol Microbiol 30*: 525-534.

Papavinasasundaram, K.G., Anderson, C., Brooks, P.C., Thomas, N.A., Movahedzadeh, F., Jenner, P.J., Colston, M.J., and Davis, E.O. (2001) Slow induction of RecA by DNA damage in *Mycobacterium tuberculosis*. *Microbiology 147*: 3271-3279.

Pitcher, R.S., Tonkin, L.M., Daley, J.M., Palmbos, P.L., Green, A.J., Velting, T.L., Brzostek, A., Korycka-Machala, M., Cresawn, S., Dziadek, J., Hatfull, G.F., Wilson, T.E., and Doherty, A.J. (2006) Mycobacteriophage exploit NHEJ to facilitate genome circularization. *Mol Cell 23*: 743-748.

Poterszman, A., Lamour, V., Egly, J.M., Moras, D., Thierry, J.C., and Poch, O. (1997) A eukaryotic XPB/ERCC3-like helicase in *Mycobacterium leprae*? *Trends Biochem Sci 22*: 418-419.

Priebe, S.D., Hadi, S.M., Greenberg, B., and Lacks, S.A. (1988) Nucleotide sequence of the *hexA* gene for DNA mismatch repair in *Streptococcus pneumoniae* and homology of *hexA* to *mutS* of *Escherichia coli* and *Salmonella typhimurium*. *J Bacteriol 170*: 190-196.

Prudhomme, M., Martin, B., Mejean, V., and Claverys, J.P. (1989) Nucleotide sequence of the *Streptococcus pneumoniae hexB* mismatch repair gene: homology of HexB to MutL of *Salmonella typhimurium* and to PMS1 of *Saccharomyces cerevisiae*. *J Bacteriol 171*: 5332-5338.

Rachman, H., Strong, M., Schaible, U., Schuchhardt, J., Hagens, K., Mollenkopf, H., Eisenberg, D., and Kaufmann, S.H. (2006a) *Mycobacterium tuberculosis* gene expression profiling within the context of protein networks. *Microbes Infect 8*: 747-757.

Rachman, H., Strong, M., Ulrichs, T., Grode, L., Schuchhardt, J., Mollenkopf, H., Kosmiadi, G.A., Eisenberg, D., and Kaufmann, S.H. (2006b) Unique transcriptome signature of *Mycobacterium tuberculosis* in pulmonary tuberculosis. *Infect Immun 74*: 1233-1242.

Rand, L. (2003) The role of DNA repair genes in *Mycobacterium tuberculosis* pathogenesis. *PhD thesis. UCL, London.*

Rand, L., Hinds, J., Springer, B., Sander, P., Buxton, R.S., and Davis, E.O. (2003) The majority of inducible DNA repair genes in *Mycobacterium tuberculosis* are induced independently of RecA. *Mol Microbiol 50*: 1031-1042.

Rangarajan, S., Woodgate, R., and Goodman, M.F. (2002) Replication restart in UV-irradiated *Escherichia coli* involving pols II, III, V, PriA, RecA and RecFOR proteins. *Mol Microbiol 43*: 617-628.

Reardon, J.T., and Sancar, A. (2005) Nucleotide excision repair. *Prog Nucleic Acid Res Mol Biol 79*: 183-235.

Reuven, N.B., Koonin, E.V., Rudd, K.E., and Deutscher, M.P. (1995) The gene for the longest known *Escherichia coli* protein is a member of helicase superfamily II. *J Bacteriol 177*: 5393-5400.

Roberts, J., and Park, J.S. (2004) Mfd, the bacterial transcription repair coupling factor: translocation, repair and termination. *Curr Opin Microbiol 7*: 120-125.

Roe, S.M., Barlow, T., Brown, T., Oram, M., Keeley, A., Tsaneva, I.R., and Pearl, L.H. (1998) Crystal structure of an octameric RuvA-Holliday junction complex. *Mol Cell 2*: 361-372.

Rupp, W.D., Sancar, A., and Sancar, G.B. (1982) Properties and regulation of the UVRABC endonuclease. *Biochimie 64*: 595-598.

Sanchez, H., Kidane, D., Castillo Cozar, M., Graumann, P.L., and Alonso, J.C. (2006) Recruitment of *Bacillus subtilis* RecN to DNA double-strand breaks in the absence of DNA end processing. *J Bacteriol 188*: 353-360.

Sander, P., Papavinasasundaram, K.G., Dick, T., Stavropoulos, E., Ellrott, K., Springer, B., Colston, M.J., and Bottger, E.C. (2001) *Mycobacterium bovis* BCG *recA* deletion mutant shows increased susceptibility to DNA-damaging agents but wild-type survival in a mouse infection model. *Infect Immun 69*: 3562-3568.

Sano, Y. (1993) Role of the *recA*-related gene adjacent to the *recA* gene in *Pseudomonas aeruginosa. J Bacteriol 175*: 2451-2454.

Sassetti, C.M., and Rubin, E.J. (2003) Genetic requirements for mycobacterial survival during infection. *PNAS 100*: 12989-12994.

Saves, I., Laneelle, M.A., Daffe, M., and Masson, J.M. (2000) Inteins invading mycobacterial RecA proteins. *FEBS Lett 480*: 221-225.

Schnappinger, D., Ehrt, S., Voskuil, M.I., Liu, Y., Mangan, J.A., Monahan, I.M., Dolganov, G., Efron, B., Butcher, P.D., Nathan, C., and Schoolnik, G.K. (2003) Transcriptional Adaptation of *Mycobacterium tuberculosis* within Macrophages: Insights into the Phagosomal Environment. *J Exp Med 198*: 693-704.

Schnarr, M., Oertel-Buchheit, P., Kazmaier, M., and Granger-Schnarr, M. (1991) DNA binding properties of the LexA repressor. *Biochimie 73*: 423-431.

Seitz, E.M., Brockman, J.P., Sandler, S.J., Clark, A.J., and Kowalczykowski, S.C. (1998) RadA protein is an archaeal RecA protein homolog that catalyzes DNA strand exchange. *Genes Dev. 12*: 1248-1253.

Sharples, G.J., Ingleston, S.M., and Lloyd, R.G. (1999) Holliday junction processing in bacteria: insights from the evolutionary conservation of RuvABC, RecG, and RusA. *J Bacteriol 181*: 5543-5550.

Sheng, D., Liu, R., Xu, Z., Singh, P., Shen, B., and Hua, Y. (2005) Dual negative regulatory mechanisms of RecX on RecA functions in radiation resistance, DNA recombination and consequent genome instability in *Deinococcus radiodurans*. *DNA Repair (Amst) 4*: 671-678.

Smith, K.C. (2004) Recombinational DNA repair: the ignored repair systems. *Bioessays 26*: 1322-1326.

Song, Y., and Sargentini, N. (1996) *Escherichia coli* DNA repair genes *radA* and *sms* are the same gene. *J. Bacteriol. 178*: 5045-5048.

Springer, B., Sander, P., Sedlacek, L., Hardt, W.D., Mizrahi, V., Schar, P., and Bottger, E.C. (2004) Lack of mismatch correction facilitates genome evolution in mycobacteria. *Mol Microbiol 53*: 1601-1609.

Stohl, E.A., Brockman, J.P., Burkle, K.L., Morimatsu, K., Kowalczykowski, S.C., and Seifert, H.S. (2003) *Escherichia coli* RecX inhibits RecA recombinase and coprotease activities *in vitro* and *in vivo*. *J Biol Chem 278*: 2278-2285.

Sukchawalit, R., Vattanaviboon, P., Utamapongchai, S., Vaughn, G., and Mongkolsuk, S. (2001) Characterization of *Xanthomonas oryzae* pv. *oryzae recX*, a gene that is required for high-level expression of *recA*. *FEMS Microbiol Lett 205*: 83-89.

Suvarnapunya, A.E., Lagasse, H.A., and Stein, M.A. (2003) The role of DNA base excision repair in the pathogenesis of *Salmonella enterica* serovar *typhimurium*. *Mol Microbiol 48*: 549-559.

Talaat, A.M., Lyons, R., Howard, S.T., and Johnston, S.A. (2004) The temporal expression profile of *Mycobacterium tuberculosis* infection in mice. *Proc Natl Acad Sci U S A 101*: 4602-4607.

Tang, M., Shen, X., Frank, E.G., O'Donnell, M., Woodgate, R., and Goodman, M.F. (1999) UmuD'(2)C is an error-prone DNA polymerase, *Escherichia coli* pol V. *Proc Natl Acad Sci U S A 96*: 8919-8924.

Taylor, A.F., and Smith, G.R. (2003) RecBCD enzyme is a DNA helicase with fast and slow motors of opposite polarity. *Nature 423*: 889-893.

Truglio, J.J., Croteau, D.L., Van Houten, B., and Kisker, C. (2006) Prokaryotic nucleotide excision repair: the UvrABC system. *Chem Rev 106*: 233-252.

Van Houten, B., Eisen, J.A., and Hanawalt, P.C. (2002) A cut above: Discovery of an alternative excision repair pathway in bacteria. *PNAS 99*: 2581-2583.

VanLoock, M.S., Yu, X., Yang, S., Galkin, V.E., Huang, H., Rajan, S.S., Anderson, W.F., Stohl, E.A., Seifert, H.S., and Egelman, E.H. (2003) Complexes of RecA with LexA and RecX differentiate between active and inactive RecA nucleoprotein filaments. *J Mol Biol 333*: 345-354.

Vaze, M.B., and Muniyappa, K. (1999) RecA protein of *Mycobacterium tuberculosis* possesses pH-dependent homologous DNA pairing and strand exchange activities: implications for allele exchange in mycobacteria. *Biochemistry 38*: 3175-3186.

Venkatesh, J., Kumar, P., Krishna, P.S.M., Manjunath, R., and Varshney, U. (2003) Importance of Uracil DNA Glycosylase in *Pseudomonas aeruginosa* and *Mycobacterium smegmatis*, G+C-rich Bacteria, in Mutation Prevention, Tolerance to Acidified Nitrite, and Endurance in Mouse Macrophages. *J. Biol. Chem. 278*: 24350-24358.

Venkatesh, R., Ganesh, N., Guhan, N., Reddy, M.S., Chandrasekhar, T., and Muniyappa, K. (2002) RecX protein abrogates ATP hydrolysis and strand exchange promoted by RecA: insights into negative regulation of homologous recombination. *Proc Natl Acad Sci U S A 99*: 12091-12096.

Vierling, S., Weber, T., Wohlleben, W., and Muth, G. (2000) Transcriptional and mutational analyses of the *Streptomyces lividans recX* gene and its interference with RecA activity. *J Bacteriol 182*: 4005-4011.

Voloshin, O.N., Vanevski, F., Khil, P.P., and Camerini-Otero, R.D. (2003) Characterization of the DNA damage-inducible helicase DinG from *Escherichia coli*. *J Biol Chem 278*: 28284-28293.

Wagner, J., Gruz, P., Kim, S.R., Yamada, M., Matsui, K., Fuchs, R.P., and Nohmi, T. (1999) The *dinB* gene encodes a novel *E. coli* DNA polymerase, DNA pol IV, involved in mutagenesis. *Mol Cell 4*: 281-286.

Walker, G.C. (1984) Mutagenesis and inducible responses to deoxyribonucleic acid damage in *Escherichia coli*. *Microbiological Reviews 48*: 60-93.

Walker, J.R., Corpina, R.A., and Goldberg, J. (2001) Structure of the Ku heterodimer bound to DNA and its implications for double-strand break repair. *Nature 412*: 607-614.

Weller, G.R., Kysela, B., Roy, R., Tonkin, L.M., Scanlan, E., Della, M., Devine, S.K., Day, J.P., Wilkinson, A., d'Adda di Fagagna, F., Devine, K.M., Bowater, R.P., Jeggo, P.A., Jackson, S.P., and Doherty, A.J. (2002) Identification of a DNA nonhomologous end-joining complex in bacteria. *Science 297*: 1686-1689.

West, S.C. (1995) Formation, translocation and resolution of Holliday junctions during homologous genetic recombination. *Philos Trans R Soc Lond B Biol Sci 347*: 21-25.

Wiederholt, C.J., Patro, J.N., Jiang, Y.L., Haraguchi, K., and Greenberg, M.M. (2005) Excision of formamidopyrimidine lesions by endonucleases III and VIII is not a major DNA repair pathway in *Escherichia coli*. *Nucleic Acids Res 33*: 3331-3338.

Wiid, I., Grundlingh, R., Bourn, W., Bradley, G., Harington, A., Hoal-van Helden, E.G., and van Helden, P. (2002) O(6)-alkylguanine-DNA alkyltransferase DNA repair in mycobacteria: pathogenic and non-pathogenic species differ. *Tuberculosis (Edinb) 82*: 45-53.

Winterling, K.W., Chafin, D., Hayes, J.J., Sun, J., Levine, A.S., Yasbin, R.E., and Woodgate, R. (1998) The *Bacillus subtilis* DinR binding site: redefinition of the consensus sequence. *J Bacteriol 180*: 2201-2211.

Zhang, Q.M., Yonei, S., and Kato, M. (1992) Multiple pathways for repair of oxidative DNA damages caused by X rays and hydrogen peroxide in *Escherichia coli*. *Radiat Res 132*: 334-338.

Zharkov, D.O., and Grollman, A.P. (1998) MutY DNA glycosylase: base release and intermediate complex formation. *Biochemistry 37*: 12384-12394.

Zhu, H., and Shuman, S. (2006) Substrate specificity and structure-function analysis of the 3'-phosphoesterase component of the bacterial NHEJ protein, DNA ligase D. *J Biol Chem 281*: 13873-13881.

In: DNA Research Trends
Editor: Marius P. Young, pp. 47-68
ISBN: 978-1-60021-646-6

Chapter II

Occurrence of DNA Endoreplication in Neurons

***L. Mola*[1]*, M. Mandrioli*[2,*]*, B. Cuoghi*[1] *and D. Sonetti*[2]**
[1]Department of Paleobiology Museum and Botanical Garden, Anatomical Museums, University of Modena and Reggio Emilia, Modena, Italy;
[2]Department of Animal Biology, University of Modena and Reggio Emilia, Modena, Modena, Italy

Abstract

DNA endoreplication has been repeatedly reported in protists, plants, arthropods, molluscs, fish and mammals. The same studies indicated that cells possessing endoreplicated genome are generally large-sized and highly metabolically active, suggesting that endoreplication could have a functional significance. Neurons are typically considered as fully differentiated, non-dividing cells containing a diploid genome. Despite this general rule, some papers questioned the validity of this statement and indicated that giant neurons in molluscs, supramedullary and hypothalamic magnocellular neurons in fish and Purkinje cells in vertebrate (predominantly mammalian) cerebellum present DNA contents larger than 2C. Quantitative microfluorimetric evaluation of DNA content in nerve cells of the gastropod molluscs *Planorbarius corneus, Aplysia californica* and *Lymnaea stagnalis* indicated that neuronal DNA contents range between 2C and 200.000C values. This increase in DNA content generally concerns whole-genome duplications, whereas in *P. corneus* endoreplication mainly applies to GC-rich sequences. The second example of endoreplicated neurons is highlighted by the large, clustered neurons located at the boundary between the medulla oblongata and spinal cord in the fish *Lophius piscatorius* and *Diodon holacanthus*. The DNA content of these neurons, evaluated by microfluorimetric methods, range from 4C in small to over 5000C in large neurons. Further experiments with AT and GC specific fluorochromes showed that the increase in DNA content is due to an amplification

* E-mail: mola.lucrezia@unimore.it.

involving GC-rich DNAs in *L. piscatorius,* whereas a whole-genome endoduplication occurs in *D. holacanthus*. Subsequent quantitative evaluation revealed that also *L. piscatorius* hypothalamic magnocellular neurons, located in the preoptic and tuberal areas, largely exceed 2C. The third example is represented by nuclei of Purkinje cells isolated from the cerebellum. These results have been debated for several years, since contrasting data are present in literature. Up till now, the dilemma remains unsolved, but it is not possible to exclude that a small percentage of Purkinje neurons contain hyperdiploid and tetraploid nuclei, due to additional DNA synthesis. In order to study this topic into more detail, we here review the available data about endoreplication in invertebrate and vertebrate giant neurons and consider the possible molecular mechanisms responsible for endoreplication. Furthermore, some possible functional significances of neuron endoploidy are discussed.

INTRODUCTION

The mechanisms involved in the control of cell cycle are generally highly conserved in eukaryotes. However, some species possess cell cycle variations and, consequently, different control mechanisms of this cycle [45].

One of the most common cell cycle variant is represented by endoreplication, in which cells increase their genomic DNA content without dividing. In particular, endoreplication is responsible for the presence of genomic DNA. In many cases the chromosome number is increased in multiples of N (the normal haploid chromosome number) bringing to polyploidy or endopolyploidy. On the other hand, in some cases, such as in the dipteran insects *Drosophila melanogaster* and *Chironomus thummi*, endoreplicated sister chromatids remain closely associated resulting into polytene chromosomes [45]. The best known example of polyteny is found in the giant salivary gland chromosomes of *D. melanogaster*, which have up to 2048 copies of the euchromatic genome neatly aligned in parallel arrays [e.g. 121].

Although endoreplication is a dismissed evolutionary peculiarity, endoduplicated genomes result widespread in protists, plants and animals (including arthropods, molluscs, and mammals). Endocycling cells become largely polyploid with C values (indicating the DNA content as a multiple of the normal haploid genome size) as high as 24000C as reported for some plants [120]. Considering that cell size is generally proportional to the amount of nuclear DNA, endoreplication constitutes an effective strategy for cellular growth and it is often found in differentiated cells that are large-sized or highly metabolically active.

Endoreplication is widespread in arthropods and it has been extensively studied in *D. melanogaster* [124]. In particular, polyploidy is a pervasive phenomenon in *Drosophila* and it occurs in both somatic tissues and derivatives of the germ-line. Examples include the nurse cells and follicle cells of the egg chamber within the ovary. Genome-wide endoreplication is observed in both the nurse cells and follicle cells and during this process endoreplication mimics a normal cell cycle that alternates S-phases and G-phases, but without any M-phase [45]. Following the cell-proliferation phase of *Drosophila* embryogenesis, many organs initiate endocycles that lack all visible aspects of mitosis [112]. These organs, which include gut, epidermis, fat body, malpighian tubules, trachea and salivary glands, continue to endocycle during larval development long after they are fully differentiated. Some adult cells, including ovarian follicle and nurse cells and the sensory neurons in the wing, also reveal endocycles. Final DNA levels in the larval cells appear to be developmentally programmed ranging from 16C to 2048C depending on cell type [45].

Endoreplication cycles were also reported for other insects and in particular for the nuclei of ovarian nurse and follicle cells of the silkmoth *Hyalophora cecropia* [32] and in the salivary glands of the dipteran *Sciara coprophila* [74]. A unique feature of the salivary gland polytene chromosomes of *S. coprophila* is that certain loci form large "DNA puffs" in late larval life, which are sites not only of intense transcription, but also of DNA amplification [56].

Cell types undergoing endoreplication were identified also in vertebrates [45]. An example is represented by megakaryocytes that, as a part of their differentiation process, become polyploid up to 128C [125]. Endopolyploidy is achieved via endomitosis and is triggered by the secreted signal thrombopoietin. The large increase in megakaryocyte size that results from polyploidy is correlated with the cell ability to bud-off large numbers of platelets. Endomitosis in megakaryocytes involves nuclear envelope breakdown and the appearance of condensed chromosomes and multipolar spindles [95]. Sister chromatids generally separate in the anaphase, but since this phase does not occur, replicated copies of chromosomes are incorporated into the same nucleus when the nuclear envelope reforms at a stage equivalent to telophase. Cytokinesis appears to be completely suppressed.

A second mammalian cell type that undergoes endocycles is the trophoblast, which contributes to the placenta, and increases its DNA contents up to more than 1000C, presumably to face an high metabolic demand [123,129]. The chromosomes of trophoblast giant cells have regions whit tightly associated replicated copies [123]. Thus, their chromosomes are polytene, although they are not aligned over their entire lengths to produce the intricate banding patterns reported in polytene chromosomes of insects.

Endoreplication has not been frequently described in neuronal cells, which are typically considered as fully differentiated, non-dividing cells containing a diploid amount of DNA. Despite this general rule, some authors questioned the validity of this notion and indicated that giant neurons in molluscs and supramedullary and hypothalamic magnocellular neurons in fish and Purkinje cells in the mammalian cerebellum present DNA contents greater than 2C.

Furthermore, some possible functional significances about neuron endoploidy are discussed.

SWITCHING FROM MITOSIS TO ENDOREPLICATION CYCLES

The endopolyploid state occurs when cells do not stop cycling when they become post-mitotic, but switch to endoreplication cycles in which repeated S-phases occur without intervening mitoses [44,45].

To date, there no report to deal with the molecular mechanisms involved in neuron endoreplication. Still, endoreplication has been dissected at the molecular level in *Drosophila* into such detail that this knowledge can be very beneficial to neuroscientists.

In *Drosophila*, the switch to endoreplication appears to be accomplished by loss of the mitotic cyclins A and B, while periodic expression of the S-phase-promoter cyclin E continues [109]. Similar observations have been made in endoreplicating maize endosperm and in several mutants in yeast [110].

The increase in cyclin-dependent kinases (cdk) activity that triggers DNA replication in yeast also makes replication origin complexes incapable of re-initiation [43]. This block of re-initiation is preserved during late S- and G2-phase by the accumulation of the mitosis-promoting cyclin-cdks and is removed when these complexes are inactivated by cyclin degradation during mitosis [43,44,110]. Thus, the absence of mitotic cyclins in endoreplicating cells may explain both the lack of mitosis and why re-initiation is no longer dependent on mitosis, whereas periodic cyclin E expression provides an explanation for how multiple rounds of DNA replication are triggered. Consistent with these ideas, inactivation of cyclin-dependent kinase 1 (cdk1) in *Drosophila* does not just arrest cells in G2, but forces them into an endoreplication cycle [44]. Moreover, although the pulses of cyclin E expression that normally preceeds the S-phase during endoreduplication are necessary and sufficient to trigger S-phases, periodic endoreplication can be inhibited by simply continuously over-expressing cyclin E.

A second set of molecules that is part of the endocycle switch on is formed by the Myc/Max/Mad network of transcription factors that play a role in a wide range of cellular processes including growth and proliferation, differentiation, apoptosis and oncogenesis [58]. Members of the Myc-family are over-expressed in various human tumors and are implicated in tumor initiation and progression. When ectopically expressed in mammalian cells, Myc proteins can induce proliferation and growth, block terminal differentiation and, in cooperation with other signals, induce transformation [58]. Conversely, over-expression of Mad-family members leads to the arrest of the cell cycle in G1 [126].

The Myc/Max/Mad network is conserved in *Drosophila* and comprises dMax and single members of the Myc and Mad families, dMyc and dMnt, respectively [54,104]. In particular, dMyc over-expression is present in endoreplicating cells resulting in dramatic increases in nuclear DNA content and cell and nucleolar size, whereas dMnt overexpression has the opposite effect. Ectopic expression of dMyc results in increased cytoplasmic and nuclear volume, as well as in enlarged nucleoli, as appears from intensified anti-fibrillarin staining [104]. Fibrillarin has been implicated as a Myc target gene in both vertebrate and *Drosophila* cells [98,104] and its increased expression is consistent with the notion that dMyc/Myc promotes ribosome biogenesis [104,119]. dMyc can post-transcriptionally increase cyclin E levels in wing discs and studies in mammalian cells suggest that Myc indirectly induces cyclin E expression [104]. Microarray analysis did not identify cyclin E as a transcriptional target of dMyc [104], suggesting that the transcriptional oscillation of cyclin E is not directly regulated by dMyc. Thus, dMyc is unable to drive endoreplication in the absence of normal cdk activity. Although the level of fibrillarin staining was not quantified, dMyc appears to drive nucleolar growth rather than DNA accumulation when co-expressed with cyclin E or p21, indicating that dMyc may be able to drive some nucleolar growth in the absence of DNA replication [104].

Experiments using BrdU staining of *Drosophila* follicle cells showed that endoreplication does not occur at whole-genome level since only four regions undergo local amplification: two of these regions, or "amplicons," are situated on the X chromosome and on chromosome 3 [22,31]. In both chromosomes, amplification is controlled by defined *cis*-acting elements that control the selective increase in chorion gene copy number [2,4]. Both amplicons have been extensively characterized during the past 20 years and much evidences suggest that the increase in chorion gene copy number is essential for the production of the large quantities of the encoded proteins that are required for the structural integrity of the eggshell [2,4]. Other genes located within the amplicons are thought to "go along for a ride" by virtue of their proximity to the chorion genes. The amplification program is exquisitely controlled and it dependends upon the repression of replication at chromosomal positions except for those regions encompassing the amplicon [2,4,22]. Amplicons might exhibit variable levels of replication in different follicle cells. For example, at late stages of maturation the anterior-dorsal follicle cells seem to incorporate more BrdU than other follicle cells of the egg chamber, indicating that there is both a spatial and temporal control in the replication process [22].

A close glance at *Drosophila* endocycles indicated that some genomic compartments are generally not endoreplicated [59,68].

These data, taken together, are very interesting because they can explain the occurrence of endoreplication also in neurons. In particular, it is possible to endoreplicate different genomic portions that contain genes whose expression is essential for the production of large quantities of proteins required for cell structural integrity and functioning. This hypothesis is strengthened by data indicating that endoreplicating cells are generally highly metabolically active. Finally, the mechanisms reported for *Drosophila* suggests that endoreplication occurs in the absence of mitosis, and this is an essential feature for neurons that are normally thought to be non-diving cells and have specific mitotic blocking mechanisms.

Giant Neurons in Gastropod Molluscs

The nervous system of invertebrates represents an intriguing and important model for neurobiological studies on neuronal evolution. Actually, invertebrates show very diversified neuronal organizations which range from the first largely unstructured nervous system of Coelenterata, forming an extensive network of some hundreds of small neurons in *Hydra*, to the most complex and large invertebrate brain found in the cephalopod mollusc *Octopus,* which reveals a sophisticated organization comparable at least to that of the lower vertebrates, showing a relative high degree of “intelligence”, learning capabilities and memory.

Giant neurons occur in many invertebrates but, apart from the special case of gastropod molluscs, where this peculiarity is almost the rule, they are not very frequent. The first reported cases of giant nerve cells concern nematodes and annelids [for a review see 30], and in particular invertebrate neurophyisiologists studied extensively the giant Retzius cells, a couple of very large neurons located in segmental ganglia of the leech *Hirudo medicinalis*, which probably represent the first biochemically characterized neurons [37,55]. Also in arthropods large neurons are present: Diptera, Odonata, Chelicerata and Arachnida are reported to possess some giant cells [30]. These neurons, identified as motor elements, usually have large-sized axonal fibres innervating main motor units and are responsible for rapid movements in response to visual stimuli as a reflex component to promote escape or aggressive behavior. At our knowledge, no reports deal with the amount of DNA in the nuclei of these neurons.

In the different classes of molluscs, we observe an increasing number of neurons and consequent complexity of CNS organization, starting from the most simply structured ones, via a "cord-like" CNS, to ganglionic formation and "cephalization" of the CNS. Ganglia are formed by aggregates of nerve cells controlling specific parts of the body. The resulting ganglia are joined by connectives and commissures that make a ring around the oesophageal tube in proximity to the buccal complex. Each ganglion has a name justifying his innervation area and consequent function, *i.e.* buccal, cerebral, pedal, pleural, parietal and visceral.

As to the ganglion cells of gastropods, Bullock and Horridge [30] described them as "extraordinary in providing the basis of a disproportionate fraction of our knowledge of neuronal cytology, because of the large size and accessibility of some of them..." and more "the largest cells are veritable giants, not only relative to others in the same animals but to the nerve cells in any group of animals and indeed to active cells in general, attaining diameters of 0.8 mm". Finally, the authors stated "They (the large and giant neurons) are especially notable for the size of the nucleus, which is commonly about two-thirds of the diameter of the cell."

In works specifically dealing with the morphology of the neuronal nuclei in Gastropods [36,60,105], the chromatin is described as being made by clumps of heterochromatin uniformly dispersed in a less deeply stained nucleoplasm. In *Planorbarius corneus,* the increase in nuclear volume is associated with both an increase in nucleoli number and in the amount of perinucleolar chromatin [75].

In the wake of this first data, Kandel and his group [46], in adopting for their research the sea-hare *Aplysia californica* (a marine opistobranch that became famous for fundamental studies on the cellular and molecular basis of behavior) made an early identification of almost all (about 30 neurons) of the very large neurons and prominent cell clusters in the abdominal ganglion. They made a morpho-functional mapping of the giant neurons and gave them an initial description: the largest neurons found are L10 and R2 reported to have a soma diameter in adults of almost 1 mm and to contain an ellipsoid nucleus with a long axis up to 500 μm, thus visible *in vivo* beneath the semitransparent connective tissue sheath, even with the naked eye! These large identifiable cells are already present in the ganglia of small juveniles, and do not change in number, whilst the number of small neurons remains increasing during the animal's life. What causes the increase in size of the large neurons and of the number of the smaller cells is not known. Probably, an intrinsic program is followed, but also a response to increased functional demands may play a role, *i.e.* a major need for innervating the growing body. Usually, small neurons are thought to be sensory cells, whereas, large neurons are motoneurons or that act as central pattern generators [122]. As Kandel [64] states, "some (neuronal) cell types, often large cells may never vary in number because they never experience demands for functional elaboration, or if they do they respond enlarging and undergoing DNA replication but not cellular replication".

The neuronal large size and the possibility to identify and map the cells in specific location within the ganglia of several gastropod species, *i.e.* the possibility to find the same individual neuron in the ganglionic ring of different specimens and sometimes even among different species, offered investigators of different neuroresearch disciplines considerable, practical advantages: single neurons could be intracellularly recorded by electrophysiologists who want to relate electrical activity with function in order to rebuilt neuronal circuitry underlying specific behavioral responses, and biochemists had the opportunity to study the chemical content of single cells in terms of neurotransmitters and other neurochemical messengers of interest in a dynamic context [for a review see 100]. The possibility to manipulate and isolate giant neurons with relative easy permitted to culture them *in vitro* in order to study under controlled conditions phenomena like neuro-degeneration/-regeneration in the presence of microglial cells [113], or to rebuild *in vitro* simple neuronal networks or circuitries [93]. The giant neurons are found in small but constant numbers in all the central ganglia at least in a variety of pulmunates and opisthobranchs, but seemingly not in prosobrachs so far examinated; few occur in the buccal, pleural, and cerebral ganglia, many in the visceral and pedal ones. Among the most studied giant neurons, in addition to those reported above from *A. californica*, are the metacerebral serotonergic neurons of *Helix* [99] and the giant dopaminergic neuron in the pedal ganglion of *P. corneus* and *Lymnaea stagnalis* [1,102].

A first study of DNA content of giant molluscan neurons was performed by Coggeshall et al. [38]. Cytophotometric analyses have shown that the DNA content in giant neurons of *A. californica* can vary, during development, from 2000C to 75000C by incremental duplications of the whole genome, thus supporting the idea that polyploidy takes place. The absence of selective duplication has been confirmed by fluorimetric analyses, showing that *Aplysia* possesses giant neurons containing an amount of DNA 200000 times higher than the haploid 1 C value present in its spermatozoa [71]. Furthermore, a quantitative analysis of large nuclei after Feulgen DNA staining revealed the presence of repeated duplication of the complete genome in other gastropods, the pulmonates snails *Helix pomatia* [78] and *L. stagnalis* [19]. The DNA content in ganglionic cells of molluscs, far in excess of the 2C and 4C normal diploid values, has been the object of a number of successive studies [for review see 26].

In contrast, in the freshwater snail *P. corneus* and the land snail *Achatina fulica,* a differential duplication of DNA, rather than a repeated whole-genome duplication, seems to take place [34,76,77,79]. Cytochemical and microfluorimetric analyses, useful in distinguishing GC-rich from AT-rich DNA strand, suggested that during the development of the CNS of *P. corneus,* the increase in DNA content and consequent nuclear volume is due to differential amplification of GC-rich DNA sequences in selected compartments of the genome. This indicates that the variations in DNA values are almost exclusively due to the extremely variable amounts of DNA and not to partially double-stranded RNAs. Moreover, the results obtained by photo-oxidation of nuclei of different dimensions show that this technique is much more destructive when applied to DNA contained in giant and large nuclei. As noted by Chase and Tolloczko [34], differential DNA endoreplication of some DNA sequences or selective gene amplification is highest during the period of animal's greatest growth and decline rapidly near the onset of sexual maturity.

Two mechanisms have been postulated to explain gene amplification: unequal mitotic crossing-over and disproportionate or saltatory replication of single tracts of DNA [67,101].

Manicardi et al. [79] have stated that the model involving unequal crossing-over is unlikely, because the neurons investigated had a suppressed mitotic phase. Instead, gene amplification in *P. corneus* is adequately explained from saltatory replication of DNA, since, in this case, additional rounds of replication will produce a variable amount of amplified sequences [67,101]. This model, already widely utilized to sustain several cases of gene amplification [106], does not require the breakage of the original gene and furthermore does not depend on mitosis. Some other theoretical considerations may explain gene amplification in molluscs. DNA amplification might concern genes involved in the synthesis of a specific neuropeptide needed for rapid biosynthesis of specific gene products; possibly, the parts of the genome playing a role in such processes are the same as those that will be differentially replicated. Alternatively, the amplified genes might be involved in a more aspecific mechanism, like, for instance, the production of a considerable amount of ribosomal DNA aimed to generally increase protein synthesis capacity. In *A. californica* the investigations by Picciotto *et al.* [103] seem to exclude an amplification of genes encoding neuropeptides. The same conclusion was reached for polyploidy in *L. stagnalis* by Smit (personal communication), who, utilizing specific cDNA probes, demonstrated that the massive production of a specific neurohormone is not related to amplification of the corresponding gene. Therefore, the most plausible explanation of the data found in *P. corneus* is that the amplified parts of the genome contain cistrons for the production of ribosomal RNA. This hypothesis is supported by the particular abundance of GC-sequences in the rDNA of many eukaryotes [111].

These data, as a whole, allow us to suggest that DNA over-replication enables a neuron to function as a substitute for a multiple neuronal system: in other words a giant molluscan neuron could be regarded as a "single cell ganglion" combining the properties of many equivalent cells. In this view, a giant neuron may assume more than one function *i.e.,* it may include a motoneuron and an interneuron, or a sensory neuron and an interneuron. This idea of multiple roles of large invertebrate neurons is supported by the fact that they produce more than one neurotransmitter. Immunohistochemical studies have shown the coexistence of two or more neuropeptides in the same cell, being produced by expression of different genes or by differential cleavage of a common protein precursor molecule [114]. Recently Sonetti et al. [115] surmised that a neuronal population in *P. corneus* including large neurons immunopositive both to anti-morphine and anti-ACTH antisera would constitute a system devoted to trauma response: initially, these cells release ACTH-like material but, thereafter, the same cells, by a "programmed" delay or by a feed-back signal system, release a morphine-like molecule to block the state of alertness and cell activation and to re-establish organism's homeostasis, as first suggested by Stefano and Scharrer [117]. Compared with other molluscan species, differences exist in the way these signaling molecules are interrelated in different molluscs; in contrast to the situation in *P. corneus* where the molecules coesist within the same, large neuron, in *Mytilus edulis* large nerve cell bodies are absent and the animal produce the molecules in different, small neurons, like in mammals. Thus, the problem of integration of different signals apparently has been solved in different ways even among molluscs, either by neuronal coexistence, or by collaborating neurons in a network. Interestingly, the "giant neuron solution" has also been used in a restricted number of cases in vertebrate, as will be explained below.

SUPRAMEDULLARY NEURONS OF FISH

Large cells, named supramedullary neurons (SN), located on the dorsal midline surface of spinal cord (under the external limiting membrane), are present in adult teleosts belonging to various orders. Credit should be given to Fritsch [47] for the discovery of these neurons in *Lophius piscatorius*, after which the studies on these peculiar nervous cells continued all over the 20th century long and they are still intriguing and represent suitable models for molecular and *in vitro* investigations.

The SN are aligned along the spinal cord in a few orders (Clupeiformes, Syngnatiformes, Scorpaeniformes, Pleuronectiformes and Perciformes), while they are clustered at the rostral end in others (Tetraodontiformes, Lophiiformes and Batrachoidiformes). These neurons have defined number and size for each species [for a complete review see 89]. Ultrastructural analysis showed that aligned SN [85,86,90] and clustered SN [39,40,96] are both engaged in a high biosynthetic activity with rapid turnover of cell organelles. For example, the presence of abundant rough endoplasmic reticulum and of a well developped Golgi apparatus, together with the heavy vesicular traffic in the cytoplasm, indicate an intense process of protein biosynthesis.

Studies on electrophysiological characteristics of SN demonstrated that these neurons, provided with an unmielinated axon, are electrically coupled [10,11,12,127] and morphological evidence was present for the existence of gap junctions among their axons [13,128]. These SN show immunoreactivity to a gastrin/CCK-like peptide [5,8,53], ACTH-like peptide [42], noradrenaline [92] and nNOS [41]. Immunohistochemical, physiological and neuroanatomical data, taken together, point out that the SN system is a component of the autonomic nervous system [89]. Its localization outside the spinal cord, the presence of noradrenaline [92] and the terminations of their axons on mucous glands in the skin [53] are providing clear support for this idea. By stimulating mucous secretion the SN may aid to protect the animal from predation, help to prevent infection from a wound that would follow a predator's attack, and to prevent the penetration of micro-organisms or parasites [127].

In the course of our investigations, it became clear that the nucleus of clustered SN in the angler fish *L. piscatorius* contained much more chromatin than other neurons in the dorsal spinal cord prompting more studies on the nucleus of these neurons.

Histochemical analysis on the nuclei of clustered SN in *L. piscatorius*, combined with computerized image analysis system, showed that the amount of DNA stained by the Feulgen's reaction is higher than that in other spinal cord neurons [6].

DNA content of *L. piscatorius* SN was then evaluated by microfluorimentric methods, using specimens ranging from 18-25 cm. Slides stained with ethidium bromide (EB) revealed neuronal DNA contents ranging from 8C in the smaller neurons (nucleus diameter 15 μm) to a maximum of 1158C in the largest ones (nucleus diameter 60 μm), demonstrating a strong correlation between nucleus size and C value. The value obtained using Acriflavine Feulgen confirmed these findings [7]. Considering that the genome size of this species is 1 pg [33], the C values obtained correspond to a minimum of 8 pg and a maximum of 1158 pg. This result represented the first remarkable finding of polyploidy in the nervous system of vertebrates.

Studying larger specimens of *L. piscatorius* (about 30 cm) revealed that the DNA content of SN can reach over 5000C, in relationship with both nucleus and animal size [108]. Utilization of AT- and GC-specific fluorochromes (DAPI and chromomycin A_3 respectively) showed that the increase in DNA contents is due to differential genome amplification involving GC-rich DNA sequences. In fact, as the nucleus size of the clustered SN increases, GC-rich sequences increase more than AT-rich ones. These data suggest that over-replication of GC-rich sequences with respect to AT-rich sequences is also correlated with the increase in animal size. Nucleolar organizing region (NOR) staining, carried out to test the involvement of NORs in the increase of nucleus area, demonstrated that each SN contains a single nucleolus [108]. Considering that both the DNA transcribing for specific mRNAs and the NORs transcribing for rRNAs, are generally rich in GC sequences, may well be that the increase in GC-rich DNA found in the clustered SN is related to increased transcriptional activity of specific genes. An other hypothesis that may explain differential endoreplication of GC-rich sequences, is that specific gene complexes would amplified to support increased production of peptides/proteins [108].

DNA amplification is a general phenomenon in clustered SN of fish, as recent cytofluorimetric research carried out in *Diodon holacanthus* (a teleost belonging to Tetraodontiformes, an order very distant from Lophiiformes) indicated DNA contents ranging from 4C (in the smaller nuclei with mean diameter of 30 μm) to more than 500C (in the larger nuclei with mean diameter of 60 μm) [91].

Ultrastructural studies performed on *D. holacanthus* SN nuclei revealed an unusually high chromatin content, rather homogeneously dispersed in the nucleoplasms, in agreement with previous cytofluorimetric data. Moreover, these ultrastructural data, combined with silver staining, suggested an intense production of ribosomal components, likely serving to satisfy high cellular demands for protein synthesis. In line with this idea, remarkably high number of nuclear pores were observed, which fits well with a high rate of nuclear activity and intense nucleus-cytoplasmic traffic [40].

In *D. holacanthus* SN nuclei the distribution of C values indicates an inter- and intra-individual variation in DNA content, which does not correspond to duplications of the whole genome, suggesting that DNA replication only applies to specific genes (endoreplication). Also, whereas the DNA content appears to be correlated with nuclear size, endoreplication occurs without preferential enrichment of either AT- or GC-rich DNA sequences [91]. Thus, the endoreplication pathways in this neuronal system are different from species to species.

Vertebrate neurons only rarely have a DNA content exceeding 2C. Therefore, on the whole, the data presented above on fish SN are remarkable in that they demonstrate DNA endoreplication in a well defined vertebrate species and identified neuronal system. The reliability of these results is based on the ease with which the SN of fish can be extracted and on their accurate identification, given their remarkable large size and their typical localization.

HYPOTHALAMIC NEUROSECRETORY NEURONS OF FISH

After its demonstration in SN of fish, the phenomenon of endoreplication was subsequently shown in other large fish neurons. Light microscopy analysis of histological sections treated with Feulgen's reaction revealed that the nucleus of the hypotalamic magnocellular neurons, located in the preoptic and tuberal areas, are intensely stained in both *L. piscatorius* and *D. holacanthus* [9]. These findings were confirmed by computerized image analysis, measuring the total nucleus area, the integrated optic density (i.o.d.) and the i.o.d. per area unit. These analysis, carried out on both hypothalamic neurosecretory neurons (preoptic and tuberal complexes) and neighbouring glial and ependimal cells, made clear that the i.o.d. per area unit is similar in all the cells in question, which indicates that the increase in the nucleus area of the neurosecretory neurons is matched by a similar increase in their Feulgen-stained DNA contents. From microfluorimetric analysis of sections treated with EB it was inferred that the hypotalamic neuronal nuclei have DNA contents that increase with increasing of nucleus size, reaching about 68C in *L. piscatorius* and about 84C in *D. holacanthus*. The relationship between the increase in nucleus size and in DNA content follows a similar pattern in both species. Observing the percentual distributions of C values in these nuclei, show that in these species, in general, the peaks did not correspond to integral multiples of the 2C diploid amount. This suggests that the increase in DNA contents is due to differential genome amplification [9].

The data on hypothalamic neurosecretory neurons indicate that, in *L. piscatorius* and *D. holacanthus,* a marked increase in DNA contents is not an exclusive feature of the SN cluster. Possibly, there is some a parallelism between the functioning of these two neuron types, sharing hypothalamic neurosecretory neurons and SN a large size and an intense biosynthetic activity of neurochemical messengers.

PURKINJE CELLS OF VERTEBRATE CEREBELLUM

A last example of endoreplication in vertebrate (especially mammalian) neurons is represented by nuclei of Purkinje cells from cerebellum and other CNS neurons, like hippocampal pyramidal neurons. The results concerned were matter of controversy for almost thirty years (from about 1965 till the early nineties). Indeed, a DNA content higher than 2C was frequently reported [14,16,17,18,20,21,23,24,25,27,28,29,57,62,63,66,69,70,73,80,83, 97,107,118], whereas, however, other Authors claimed a 2C DNA amount in these cells [35,48,49,51,52,81,82,84,88,94].

Purkinje cells of the cerebellum have been particularly studied because, as in the case of fish SN, their distinct morphology and easy and reliable identification. Many studies suggested that Purkinje cells DNA content exceed 2C and that they are tetraploid or even octoploid in several mammals, including human [69,70,80], cat [61,62] and rat [21,73,97,107]. Other studies questioned the validity of these findings and indicated that Purkinje cell nuclei have normal diploid DNA contents, e.g. in rat [51,81,94], mouse [35,81], rabbit and Guinea pig [81] and human [82,84]. Differences in methodological approaches and measurement errors could be responsible for these conflicting data [52,83]. Indeed, some

Authors proposed that the special characteristics of Purkinje cells, *i. e.* a large nucleus, together with dense cytoplasm, would introduce cytophotometry artefacts [50].

For these reasons, the importance of making appropriate corrections considering non specific factors for reliable measurements of DNA content have been stressed [81]. A critical analysis of various cytophotometric, radioautographic and biochemical approaches to the problem of CNS neuron polyploidization revealed potentially serious flaws in many of them, rendering virtually impossible interpretation of the numerous contradictory results in the literature [118].

In spite of these criticisms, a cytophotometric study of DNA content performed in Purkinje cells of rats, cats, chicken and humans revealed that in a certain number of cells the amount of DNA ranges between the diploid and tetraploid level. The incidence of tetraploid Purkinje cells varies among different species and large interindividual differences were observed. In rats, for instance, the incidence of these cells varies from 1 to 23 %. Densitometric analysis of the distribution of nuclear chromatin showed that tetraploid Purkinje cells are richer in condensed chromatin, especially near the nucleolus, where apparently the surplus of DNA is situated. It has been proposed that DNA hyperdiploidy arises as a result of either an incomplete S-phase in some immature Purkinje cell precursors or the amplification of DNA sequences, especially those localized in the nucleolar region [27].

Moreover, although cytofluorimetric methods encompass an intrinsic error of about 10-15%, which may interfere with evaluations of small differences in DNA content, the existence of some “extra-DNA” (in the range 2.2C - 2.5C) in neuron nuclei in the human CNS (pre- and post-central girus, cerebellum cortex and spinal cord) was postulated [20]. The Authors suggested that this hyperploidy would play an important role in the functional activity of the CNS.

Studies on Purkinje neurons DNA-contents in several vertebrate species such as rainbow trout, Amazon molly fish, salamander, mouse, rat, rabbit, cat, dog, monkey and human, have shown a low incidence of Purkinje cell nuclei with interclass DNA amounts. Also a low frequency (1-7%) of Purkinje cell nuclei with 4C DNA levels was observed in all species examined, except salamander and rabbit [72,87].

From comparing two stages of eel life cycle (yellow eel and silver eel) characterised by a different degree of swimming activity, a relationship between the heterogeneity of Purkinje cell chromatin and afferent pathways to the cerebellum areas has been suggested [16]. DNA contents of Purkinje neurons have also been related to hibernation in frog [3] and hedgehog [18]. These studies demonstrated that nuclear heterogeneity (diploid-hyperdiploid) in Purkinje cell population reflects differences in metabolic states. Furthermore, a link between the susceptibility of cerebellar cells to hypoxia and DNA amount was recently suggested [65].

After transplanting the cerebellum of rat embryos into the sensomotor cortex of adult rats, DNA content in granule and Purkinje cells of the transplant were determined cytophotometrically. Granule cells appeared to remain diploid, whereas about 3% Purkinje cells contain hyperdiploid and tetraploid nuclei, similar to their contents in the adult cerebellum. Therefore, these data indicate that cell hyperploidy does not depend on the functional state of the cell, but is intrinsic property of the tissue [87].

More recently, the discussion about Purkinje neurons has faded somewhat, but the dilemma as to the occurrence and significance of polyploidy of Purkinje cells remains unsolved. Obviously, it is not possible to exclude that a small percentage of Purkinje neurons contains hyperdiploid and tetraploid nuclei, due to additional DNA synthesis, the significance of which still remain unclear. In any case, this phenomenon may not be comparable with the situation in SN and hypothalamic neurosecretory neurons in which the DNA amount reaches such high values that is impossible to deny the occurrence of endoreplication.

CONCLUSION

Given its widespread occurrence in evolutionary distant species and in highly different cell types, it is not easy to provide an unique explanation for the role endoduplication plays in neurons. Nevertheless, at least two generalizations can be made. Firstly, endoreplication seems to be typical for large or metabolically active cells, suggesting that endocycling is an effective tool to allow cells to increase their mass or metabolic output. Secondly, since endoreplication permits growth without periodic rearrangements of cytoskeletal elements or cell-cell contacts, as happens in mitosis, it may cause less disruption of the structure of a highly differentiated cell than mitotic proliferation. Together, these properties make endocycle an advantageous strategy for cells and tissues that, althought strongly differentiated, still must continue to grow. These features are extremely interesting for neurons that are fully differentiated and non-dividing cells, since they could improve and increase their functional plasticity without disrupting the nervous system structure.

The analysis of endoreplication phenomena in neurons allows us to obtain some insight into the frequency of neuronal endocycling. First, it should be emphasize that giant or large-sized neurons, althought they have been found in a wide variety of invertebrate and vertebrate species, are generally not very frequent. In fact, if the occurrence of endoreplication is considered from an evolutionary perspective, it seems to have been "reinvented" several times during evolution. In particular, the presence of a single polyploid cell may be considered as a substitute for the multiple cell system: in other words a giant or large neuron could be regarded as a "single cell ganglion" combining the properties of many equivalent cells. This evolutionary choice could be useful from a functional point of view since it allows for an increase in cell functionality without affecting tissue structure or organization. However, the presence of a unique multifunctional and highly active neuron could represent an important weak point of the nervous system as well, since makes it more susceptible to damage compared to a multiple cell system that could be reorganized after damage to a single cell. Moreover, a system made of several cells could be more efficient than a single large neuron in term of assuring integration of the afferent signals and successive intracellular transduction of these signals into one, final output.

A second common element among polyploid or endopolyploid neurons appears from the fact that neuronal ploidy is correlated with animal size. This feature is not surprising since the increase in animal size could involve an increased demand for neuronal functionality that can be met by enhancing neuronal ability in synthesizing peptides or other cellular components.

This notion is well in line with our hypothesis that endocycling is an effective tool to increase cellular metabolic activity.

DNA endoreplication can be obtained through rounds of amplification of specific gene sequences or by generalized whole-genome amplifications. In this regard, a discrepancy exists as to the data reported for neurons, since in the molluscs *A. californica, H. pomatia* and *L. stagnalis*, neuronal DNA contents vary by incremental duplications of the whole genome, whereas in the land snail *A. fulica,* in the planorbid *P. corneus* and in all the described endoploid vertebrate neurons a differential duplication of specific DNA sequences rather than a repeated duplication of the whole genome is reported. This discrepancy could be due to the absence of clustering in amplicons of those genes whose amplification is necessary to satisfy neuron functional requirements in *A. californica, H. pomatia* and *L. stagnalis*. The presence of a DNA endoreplication consisting in rounds of amplification of specific gene instead of generalized whole-genome amplifications represents an evolutionary advantage as the amplification of unnecessary DNA sequences and/or “trash DNAs” can be avoided. Furthermore, the possibility to amplify different genes in each cell type will increase the functionality of endoreplication since only those genes will be multiplied that are essential for the production of large quantities of proteins required for cell structural maintenance or functioning. Cells that exert the same functions will therefore amplify essentially the same genes in view of the same functional needs.

A further discrepancy seem to relate to the observed differences in the amplified genomic compartments since, for example *L. piscatorius* and *P. corneus* amplify predominantly GC-rich DNAs, whereas *D. holacanthus* and some molluscs do not show any preferential AT or GC-rich DNA amplification. However, these results do not necessarily be controversial, as amplicons can consist of genes whose proteins are essential for cells and DNA sequences that "go along for a ride" by virtue of their proximity to the functional important genes. This endoreplication feature suggests that the differential amplification that occurs for example in the fish *L. piscatorius* and *D. holacanthus* does not reflect different functions of SNs but only a difference in the type of genes that are amplified passively because of their coupling to functional important ones. This result could be simply due to difference in genome organization occurring in the two studied fish species that are phylogenetically distant. A similar hypothesis could also explain the data observed in molluscs.

Despite the amount of available results, a number of substantial and intriguing questions still remain. For example, what and why genetic programs mediate the switch from mitotic to endocycles? What programs define which cell types will undergo endoreplication? How frequently did novel mechanisms for endocycling arise during evolution, and how extreme are the variations in regulation? Answers to these questions promise to provide insights that pertain not only to endoreplication, but also to the mechanisms used in proliferative cycles and growth control.

Ackowledgements

The Authors wish to thank Prof. E. W. Roubos, Head Department Cellular Animal Physiology, Faculty of Science Radboud University Nijmegen (The Netherlands) for critical reading and for corrections to the manuscripts.

REFERENCES

[1] Audesirk, G. (1985). Amine-containing neurons in the brain of *Lymnaea stagnalis*: distribution and effects of precursors. *Comp. Biochem. Physiol. A., 81*, 359-65.

[2] Austin, R. J., Orr-Weaver, T. L. & Bell, S. P. (1999). *Drosophila* ORC specifically binds to ACE3, an origin of DNA replication control element. *Genes Dev., 13*, 2639-2649.

[3] Barni, S., Bernocchi, G. & Biggiogera, M. (1983). Chromatin organization in frog Purkinje neurons during the annual cycle: cytochemical and ultrastructural studies. *Bas. Appl. Histochem., 27*, 129-140.

[4] Beall, E. L., Manak, J. R., Zhou, S., Bell, M., Lipsick, J. S. & Botchan, M. R. (2002). Role for a *Drosophila* Myb-containing protein complex in site-specific DNA replication. *Nature, 420*, 833–837.

[5] Benedetti, I. & Mola, L. (1988). Survey of neuropeptide-like immunoreactivity in supramedullary neurons of *Coris julis* (L.). *Brain Res., 449*, 373-376.

[6] Benedetti, I. & Mola, L. (1991). Preliminary findings on the nucleus of large neurons in *Lophius piscatorius* L. (Osteichthyes, Lophiiformes). *Eur. J. Appl. Histochem., 35*, 245-248.

[7] Benedetti, I., Manicardi, G. C. & Mola, L. (1993). Cytofluorimetric determination of DNA content in large neurons of *Lophius piscatorius* L. (Osteichthyes, Lophiiformes). *Eur. J. Histochem., 37*, 91-95.

[8] Benedetti. I., Mola, L., Marini, M. & Calzolari, C. (1993). Neuroanatomical and immunohistochemical studies on the dorsal neurons in the spinal cord of *Trigla lucerna* L. and *Scorpaena porcus* L. (Scorpaeniformes). *Ann. Anat., 175*, 77-80.

[9] Benedetti, I., Sassi, D., Mescoli, G. & Manicardi, G. C. (1999). High values of DNA content in the hypotalamic neurons of *Lophius piscatorius* and *Diodon holacanthus* (Osteichthyes). *Caryologia, 52*, 141-146.

[10] Bennett, M. L. V., Crain, S. M. & Grundfest, H. (1959). Electrophysiology of supramedullary neurons in *Sphaeroides maculatus*. I. Orthodromic and antidromic responses. *J. Gen. Physiol., 43,* 159-188.

[11] Bennett, M. L. V., Crain, S. M. & Grundfest, H. (1959). Electrophysiology of supramedullary neurons in *Sphaeroides maculatus.* II. Properties of the electrically excitable membrane. *J. Gen. Physiol., 43,* 189-219.

[12] Bennett, M. L. V., Crain, S. M. & Grundfest, H. (1959). Electrophysiology of supramedullary neurons in *Sphaeroides maculatus.* III. Organization of the supramedullary neurons. *J. Gen. Physiol., 43,* 221-250.

[13] Bennett, M. L. V., Nakajima, Y. & Pappas, G. D. (1967). Physiology and ultrastructure of electrotonic junctions. I. Supramedullary neurons. *J. Neurophysiol., 30*, 161-179.

[14] Bernocchi, G. (1983). Feulgen-DNA patterns of Purkinje cell population in vertebrates with different cerebellar cytoarchitectonics. *J. Hirnforsch., 24,* 35-42.

[15] Bernocchi, G. (1985). Cytochemical variations in Purkinje neuron nuclei of cerebellar areas with different afferent systems in *Rana esculenta*. Comparison between activity and hibernation. *J. Hirnforsch, 26,* 659-665.

[16] Bernocchi, G. & Barni, S. (1985). On the heterogeneity of Purkinje neurons in vertebrates. Cytochemical and morphological studies of chromatin during eel (*Anguilla anguilla* L.) life cycle. *J. Hirnforsch., 26*, 227-235.

[17] Bernocchi, G. & Scherini, E. (1981). Maternal protein malnutrition and post-natal cerebellar histogenesis in the rat. Effects on feulgen-DNA content of Purkinje cell population. *Cell Mol. Biol. Incl. Cyto. Enzymol., 27*, 253-261.

[18] Bernocchi, G., Barni, S. & Scherini, E. (1986). The annual cycle of *Erinaceus europaeus* L. as a model for a further study of cytochemical heterogeneity in Purkinje neuron nuclei. *Neuroscience, 17*, 427-437.

[19] Boer, H. H., Groot, C., De Jong-Brink, M. & Cornelisse, C. J. (1977). Polyploidy in the freshwater snail *Lymnaea stagnalis* (Gastropoda Pulmonata), A cytophotometric analysis of the DNA in neurons and some other cell types. *Neth. J. Zool., 27*, 245-252.

[20] Bohm, N., Kroner, B. & Kaiser, E. (1981). Cytophotometric evidence of non-S-phase extra-DNA in human neuronal nuclei. *Cell Tissue Kinet., 14*, 433-444.

[21] Bohn, R. C. & Mitchell, R. B. (1976). Cytophotometric identification of tetraploid Purkinje cells in young and aged rats. *J. Neurobiol., 7*, 255-258.

[22] Botchan, M. & Levine, M. A. (2004). Genome analysis of endoreplication in the *Drosophila* ovary. *Dev. Cell, 6*, 4-5.

[23] Bregnard, A., Knusel, A. & Kuenzle, C. C. (1975). Are all the neuronal nuclei polyploid ? *Histochemistry, 43*, 59-62.

[24] Bregnard, A., Kuenzle, C. C. & Ruch, F. (1977). Cytophotometric and autoradiographic evidence for post-natal DNA synthesis in neurons of the rat cerebral cortex. *Exp. Cell Res., 107*, 151-157.

[25] Bregnard, A., Ruch, F., Lutz, H. & Kuenzle, C. C. (1979). Histones and DNA increase synchronously in neurons during early postnatal development of the rat forebrain cortex. *Histochemistry, 61*, 271-279.

[26] Brodsky, V. Y. & Uryvaeva, I. V. (1985). *Genome multiplication in growth and development,* Cambridge, UK: Cambridge University Press.

[27] Brodsky, V. Y., Marshak, T. L., Mares, V., Lodin, Z., Fulop, Z. & Lebedev, E. A. (1979). Constancy and variability in the content of DNA in cerebellar Purkinje cell nuclei. A cytophotometric study. *Histochemistry, 59*, 233-248.

[28] Brodsky, V. Y., Marshak, T. L., Mikeladze, Z. A., Moskovkin, G. N. & Sadykova, M. K. (1984). DNA synthesis in the Purkinje neurons. *Bas. Appl. Histochem., 28*, 187-194.

[29] Brodski, V. Y., Sokolova, G. A. & Manakova, T. E. (1971). Multiple increase of DNA in the Purkinje cells of cerebellum in ontogenesis of rats. *Ontogeny,* 2, 33-36.

[30] Bullock, T. H., & Horridge, G. A. (1965). *Structure and function in the nervous system of invertebrates.* vol. II. San Francisco, USA: Freeman & Co.

[31] Calvi, B. R., Lilly, M. A. & Spradling, A. C. (1998). Cell cycle control of chorion gene amplification. *Genes Dev., 12*, 734-744.

[32] Cardoen, J., Watson, C., Deloof, A., Spencer, A. & Berry J. (2005) Polyploidy in the nuclei of ovarian nurse and follicle cells of the silk moth, *Hyalophora cecropia. Arch. Insect Biochem. Physiol.*, 15, 93- 100.

[33] Cavalier-Smith, T. (1991). Coevolution in vertebrate genome, cell and nuclear size. In G. Ghiara (Ed.), *Selected Symposia and monographs: Symposium of the evolution of terrestrial vertebrates* (pp. 51-86). Modena, Italy: Mucchi.

[34] Chase, R. & Tolloczko, B. (1987). Evidence for differential DNA endoreplication during the development of a molluscan brain. *J. Neurobiol., 18*, 395-406.

[35] Cohen, J., Mares, V. & Lodin, Z. (1973). DNA content of purified preparations of mouse Purkinje neurons isolated by a velocity sedimentation technique. *J. Neurochem., 20*, 651-657.

[36] Coggeshall, R. E. (1967). A light and electron microscope study of the abdominal ganglion of *Aplysia californica. J. Neurophysiol., 30*, 1263-1287.

[37] Coggeshall, R. E. (1972). Autoradiographic and chemical location of 5-Hydroxytryptamine in identified neurons in the leech. *Anat. Rec., 172,* 489-498.

[38] Coggeshall, R. E, Yaksta, B. A. & Swartz, F. J. (1970). A cytophotometric analysis of the DNA in the nucleus of the giant cell, R-2, in *Aplysia. Chromosoma, 32,* 205-212.

[39] Cuoghi, B. (2001). Glial cells: basic components of clusters of supramedullary neurons in pufferfishes. *J. Neurocytol., 30*, 503-513.

[40] Cuoghi, B. & Marini, M. (2001). Ultrastructural and cytochemical features of the supramedullary neurons of the pufferfish *Diodon holacanthus* (L.) (Osteichthyes). *Tissue Cell, 33*, 491-499.

[41] Cuoghi, B., Marini, M. & Mola, L. (2002). Histochemical and immunohistochemical localization of nitric oxide synthase in the supramedullary neurons of the pufferfish *Tetraodon fluviatilis. Brain Res., 938*, 3-6.

[42] Cuoghi, B., Blasiol, L. & Sabatini, M. A. (2003). ACTH occurrence in teleosts supramedullary neuron clusters: a neuron-glial common language ? *Gen. Comp. Endocrinol., 132*, 88-95.

[43] Dahmann, C., Diffley, J.F. & Nasmyth, K.A. (1995). S-phase-promoting cyclin-dependent kinases prevent re-replication by inhibiting the transition of replication origins to a pre-replicative state. *Curr. Biol., 5*, 1257-1269.

[44] Edgar, B. A. & Lehner, C. F. (1996). Developmental Control of Cell Cycle Regulators: A Fly's Perspective. *Science,* 274, 1646-1652.

[45] Edgar, B. A. & Orr-Weaver, T. L. (2001). Endoreplication Cell Cycles: More for Less. *Cell, 105,* 297-306.

[46] Frazier, W. T., Kandel E. R., Kupfermann, I. & Coggeshall R. E. (1967). Morphological and functional properties of identified neurons in the abdominal ganglion of *Aplysia californica. J. Neurophysiol., 30*, 1288-1351.

[47] Fritsch, G. (1886). Ueber einige bemerkenswerhe elemente des central nerven system von *Lophius piscatorius. Arch. Mikr. NAT. Entw.-Meech., 27*, 13-31.

[48] Fujita, S. (1973). DNA cytofluorimetry on large and small cell nuclei stained with pararosaniline Feulgen. *Histochimie, 36*, 193-199.

[49] Fujita, S. (1974). DNA constancy in neurons of human cerebellum and spinal cord as revealed by Feulgen cytofluorimetry and cytofluorimetry. *J. Comp. Neurol., 155*, 195-202.

[50] Fujita, S., Fukuda, M., Kitamura, T. & Yoshida, S. (1972). Two wave-length-scanning method in Feulgen cytophotometry. *Acta Histochem. Cytochem., 5*, 146-152.

[51] Fujita, S., Hattori, T., Fukuda, M. & Kitamura, T. (1974). DNA contents in Purkinje cells and inner granule neurons in developing rat certebellum. *Dev. Growth Diff., 16*, 205-212.

[52] Fukuda, M., Bohm, N. & Fujita, S. (1978). DNA constancy in neurons of the rat and human cerebellum. *Progr. Histochem. Cytochem., 11*, 20-24.

[53] Funakoshi, K., Katoda, T., Atobe, Y., Nakamo, M., Goris, R. C. & Kishida, R. (1998). Gastrin/CCK-ergic innervation of cutaneous mucous gland by the supramedullary cells of the puffer fish *Takifugu niphobles. Neurosci. Lett., 258*, 171-174.

[54] Gallant, P., Shiio, Y., Cheng, P. F., Parkhurst, S. M. & Eisenman, R. N. (1996). Myc and Max homologs in *Drosophila. Science, 274*, 1523-1527.

[55] Gaskell, J. F. (1919). Adrenalin in annelids. *J. Physiol., 2*, 73-85.

[56] Gerbi, S. A. & Urnov, F. D. (1996). Differential DNA replication in insects, In M. L. DePamphilis (Ed.), *DNA replication in eukaryotic cells* (pp. 947-969). Cold Spring Harbor, N.Y, USA: Cold Spring Harbor Laboratory Press.

[57] Giacometti, S., Scherini, E. & Bernocchi, G. (1986). Nuclear heterogeneity of the Purkinje neuron population in the cerebellum of the reptiles. *Boll. Zool., 53,* 359-364.

[58] Grandori, C., Cowley, S. M., James, L. P. & Eisenman, R. N. (2000). The Myc/Max/Mad network and the transcriptional control of cell behavior. *Ann. Rev. Cell Dev. Biol., 16*, 653-699.

[59] Hammond, M. P. & Laird, C. D. (1985). Chromosome structure and DNA replication in nurse and follicle cells of *Drosophila melanogaster*. *Chromosoma, 91*, 267-278.

[60] Heinz, E. (1944). Kleine Beitiage zur Zellenlehre II. Uber die Riesenkerne der Schneken und Asseln. *Rev. Suisse Zool., 51*, 402-404.

[61] Herman, C. J. & Lapham, L. W. (1968). DNA content of neurones in the cat hippocampus. *Science, 160*, 537.

[62] Herman, C. J. & Lapham, L. W. (1969). Neuronal polyploidy and nuclear volume in cat central nervous system. *Brain Res., 15,* 35-48.

[63] Hobi, R., Studer, M., Ruch, F. & Kuenzle, C. C. (1984). The DNA content of cerebral cortex neurons determinations by cytophotometry and high performance liquid chromatography. *Brain Res., 305*, 209-219.

[64] Kandel, E. R. (1976). *Cellular Basis of Behavior*. San Francisco, USA: Freeman & Co., p.221.

[65] Kowalska, E. & Marcinkowski, T. (1998). The amount of DNA in cells of the granular layer of the cerebellum and their susceptibility to hypoxia. *Med. Hypotheses, 51*, 443-444.

[66] Kuenzle, C. C., Bregrard, A., Hubschner, U. & Ruch, F. (1978). Extra DNA in forebrain neurons. *Exp. Cell Res., 113*, 151-160.

[67] Kunz, W. (1984). How genes are amplified in germ cells and somatic cells. In W. Engels (Ed.), *Advances in Invertebrate Reproduction*. (vol. III, pp. 65-71). New York, USA: Elsevier Press

[68] Laird, C. D. (1980). Structural paradox of polytene chromosomes. *Cell, 22*, 869-874.

[69] Lapham, L. W. (1965). The tetraploid DNA content of normal human Purkinje cells and its development during the perinatal period. A quantitative cytochemical study. *Excerpta Med. (Amst.) Cong. Ser., 100*, 445-449.

[70] Lapham, L. W. (1968). Tetraploid DNA content of purkinje neurones of human cerebellar cortex. *Science, 159*, 310-312.

[71] Lasek, R. J. & Dower, W. J. (1971). *Aplysia californica*: analysis of nuclear DNA in individual nuclei of giant neurons. *Science, 172*, 278-280.

[72] Lee, G. M., Rasch, E.M. & Thornthwaite, J. T. (1984). Cytophotometric comparisons of DNA levels in neuronal and glial cells of the cerebellum: a comparative study. *Cell Biochem. Funct., 2*, 225-236.

[73] Lentz, R. D. & Lapham, L. W. (1969). A quantitative cytochemical study of the DNA content of neurons of rat cerebellar cortex. *J. Neurochem., 16*, 379-384.

[74] Liang, C., Spitzer, J. D., Smith, H. S. & Gerbi, S. A. (1993). Replication initiates at a confined region during DNA amplification in *Sciara* DNA puff II/9A. *Genes Dev., 7,* 1072-1084.

[75] Lombardo, F. & Sonetti, D. (1977). The neuronal nuclei in the central ganglia of Planorbid snails. *The Nucleus, 20*, 271-277.

[76] Lombardo, F. & Sonetti, D. (1983). Amplification and under-replication of repetitive DNAs in the neuronal nuclei of *Planorbis corneus*. *Neurosci. Lett. Suppl., 14,* 226.

[77] Lombardo, F., Baraldi, E. & Sonetti, D. (1980). Differential staining and fluorescence of chromatin in population of neuronal nuclei from *Planorbis corneus*. *The Nucleus, 23*, 30-36.

[78] Manfredi Romanini, M. G., Fraschini, A. & Bernocchi, G. (1972). DNA content and nuclear area in the neurons of the cerebral ganglion in *Helix pomatia*. *Ann. Histochem., 18*, 49-58.

[79] Manicardi, G. C., Bizzaro, D., Sonetti, D., Lombardo, F. & Bianchi, U. (1992). Amplification of GC-rich DNAs in neuronal nuclei of *Planorbarius corneus* (L.) (Mollusca, Pulmonata). *Eur. J. Histochem., 36*, 303-309.

[80] Mann, D. M. A. & Yates, P. O. (1973). Polyploidy in the human nervous system. Part I. The DNA content of neurones and glia of the cerebellum. *J. Neurol. Sci., 18,* 183-196.

[81] Mann, D. M. A, Yates, P. O & Barton, C. M. (1978). The DNA content of Purkinje cells in mammals. *J. Comp. Neurol., 180*, 345-347.

[82] Manuelidis, L. & Manuelidis, E. E. (1974). On the DNA content of cerebellar Purkinje cells *in vivo* and *in vitro*. *Exp. Neurol.*, 43, 192-206.

[83] Mares, V. & van-der-Ploeg, M. (1980). Cytophotometric re-investigation of DNA content in Purkinje cells of the rat cerebellum. *Histochemistry, 69*, 161-167.

[84] Mares, V., Lodin, Z. & Sacha, J. (1973). A cytochemical autoradiographic study of nuclear DNA in mouse Purkinje cells. *Brain Res., 53*, 273-289.

[85] Marini, M. & Benedetti, I. (1992). The Rohon-Beard cells and supramedullary neurons in Teleosts. In I. Benedetti, B. Bertolini & E. Capanna (Eds.), *Selected Symposia and monographs: Neurology today* (pp. 217-236). Modena, Italy: Mucchi.

[86] Marini, M., Benedetti, I. & Franchini, (1984). Ultrastructural features of supramedullary neurons in *Crenilabrus quinquemaculatus*. *Z. mikrosk-anat. Forsch, Leipzig, 98*, 63-71.

[87] Marshak, T. L., Mares, V. & Brodsky, V. Y. (1985). An attempt to influence DNA content in postmitotic Purkinje cells of the cerebellum. *Acta Histochem., 76*, 193-200.

[88] McIlwain, D. L. & Capps-Covey, P. (1976). The nuclear DNA content of large ventral spinal neurons. *J. Neurochem., 27*, 109-112.

[89] Mola, L. & Cuoghi, B. (2004). The supramedullary neurons of fish: present status and goals for the future. *Brain Res. Bull., 64*, 195-204.

[90] Mola, L., Marini, M. & Benedetti, I. (1992). Acid phosphatase activity in the supramedullary neurons of *Coris julis* (L.). *Eur. J. Histochem., 36*, 233-236.

[91] Mola, L., Cuoghi, B., Mandrioli, M. & Marini, M. (2001). DNA endoreplication in the clustered supramedullary neurons of the pufferfish *Diodon holacanthus* L. (Osteichthyes). *Histochem. J., 33*, 59-63.

[92] Mola, L., Sassi, D. & Cuoghi, B. (2002). The supramedullary cells of the teleost *Coris julis* (L.): a noradrenergic neuronal system. *Eur. J. Histochem., 46*, 329-332.

[93] Montarolo, P. G., Kandel, E. R. & Schacher, S. (1988). Long-term heterosynaptic inhibition in *Aplysia. Nature, 12,* 333, 171-174.

[94] Morselt, A. F. W., Braakman, D. J. & James, J. (1972). Feulgen-DNA and fast green histone estimation in individual cell nuclei of cerebellum of young and old rats. *Acta Histochem., 43*, 281-286.

[95] Nagata, Y., Muro, Y. & Todokoro, K. (1997). Thrombopoietin-induced polyploidization of bone marrow megakaryocytes is due to a unique regulatory mechanism in late mitosis. *J. Cell Biol., 139*, 449-457.

[96] Nakajima, Y., Pappas, G. D. & Bennett, M. V. L. (1965). The fine structure of the supramedullary neurons of the puffer with special reference to endocellular and pericellular capillaries. *Am. J. Anat., 116*, 471-492.

[97] Novakovà, V., Sandritter, W. & Schlueter, G. (1970). DNA content of neurons in rat central nervous system. *Exp. Cell Res., 60,* 454-456.

[98] Orian, A., van Steensel, B., Delrow, J., Bussemaker, H. J., Li, L., Sawado, T., Williams, E., Loo, L. W., Cowley, S. M., Yost, C. (2003). Genomic binding by the *Drosophila* Myc, Max, Mad/Mnt transcription factor network. *Genes Dev. 17*, 1101–1114.

[99] Osborne, N. N. (1972). The *in vivo* synthesis of serotonin in an identified serotonin-containing neuron of *Helix pomatia. Int. J. Neurosci., 3,* 215-219.

[100] Osborne N. N. (1978). *Biochemistry of characterised neuron*. Oxford, UK: Pergamon Press.

[101] Osheim, Y. N. & Miller, O. L. Jr. (1983). Novel amplification and transcriptional activity of chorion genes in *Drosophila melanogaster* follicle cells. *Cell, 33,* 543-553.

[102] Pentreath, V. W., Berry, M. S. & Cottrell, G. A. (1974). Anatomy of the giant dopamine-containing neurone in the left pedal ganglion of *Planorbis corneus*. *Cell Tissue Res. 151,* 369-384.

[103] Picciotto, M. R., Johnshon, R. & Schellers, R. M. (1986). Neuropeptide genes are not selectively amplified in *Aplysia* polyploid neurons. *Neurol. Neurobiol., 20,* 429-434.

[104] Pierce, S. B., Yost B., Britton, J. S., Flynn, E. M., Edgar, B. A. & Eisenman, E. N. (2004). dMyc is required for larval growth and endoreplication in *Drosophila. Development, 131*, 2317-2327.

[105] Quattrini, D. (1950). Particolarità strutturali del nucleo di *Helix vermiculata* Muller. *Rend. Acc. Naz. Lincei (ser. VIII), 8*, 149-152.

[106] Roberts, J. M., Buck, L. B. & Axel, R. (1983). A structure for amplified DNA, *Cell, 33*, 53-63.

[107] Sandritter, W., Novakovà, V., Pilny, J. & Kiefer, G. (1967). Cytophotometrische Messungen des Nuklein-Saure und Proteingehaltes von Ganglienzellen der Ratte wahrend der postnatalen Entwicklung und im Alter. *Z. Zellforsch., 80*, 145-152.

[108] Sassi, D., Manicardi, G. C., Mola, L. & Benedetti, I. (1995). Cytofluorimetric evidence for differential genome endoreplication in the cluster neurons of *Lophius piscatorius* L. (Osteichthyes, Lophiiformes). *Eur. J. Histochem., 39*, 117-126.

[109] Sauer, K., Knoblich, J.A., Richardson, H. & Lehner, C.F. (1995). Distinct modes of cyclin E/cdc2c kinase regulation and S-phase control in mitotic and endoreduplication cycles of *Drosophila* embryogenesis. *Genes Dev*., 9, 1327-1339.

[110] Schwob, E., Bohm, T., Mendenhall, M.D. & Nasmyth, K. (1994). The B-type cyclin kinase inhibitor p40SIC1 controls the G1 to S transition in *S. cerevisiae*. *Cell*, 79, 233-244.

[111] Sinclair, J. H. & Brown, D. D. (1971). Retention of common nucleotide sequences in the ribosomal deoxyribonucleic acid of eukaryotes and some of their physical characteristics. *Biochemistry, 10*, 2761-2769.

[112] Smith, A.V. & Orr-Weaver, T.L. (1991). The regulation of the cell cycle during *Drosophila* embryogenesis: the transition to polyteny. *Development, 112*, 997-1008.

[113] Sonetti, D. & Peruzzi, E. (2004). Neuron-microglia communication in the CNS of the freshwater snail *Planorbarius corneus*. *Acta Biol. Hung., 55,* 273-285.

[114] Sonetti, D., Lusvardi, C. & Fasolo, A. (1990). Immunohistochemical localization of some vertebrate neuropeptides (SP, NPY, CGRP, CCK8) in the central nervous system of the freshwater snail *Planorbarius corneus. Cell Tissue Res., 260*, 435-448.

[115] Sonetti, D., Peruzzi, E. & Stefano, G. B. (2005). Endogenous morphine and ACTH association in neural tissues. *Med. Sci. Monit. 11*, MS22-30.

[116] Spradling, A.C. (1981). The organization and amplification of two chromosomal domains containing *Drosophila* chorion genes. *Cell, 27*, 193-201.

[117] Stefano, G. B. & Sharrer, B. (1994). Endogenous morphine and related opiates, a new class of chemical messengers. *Adv. Neuroimmunol., 4*, 57-68.

[118] Swartz, F. J. & Bhatnagar, K. P. (1981). Are CNS neurons polyploid? A critical analysis based upon cytophotometric study of the DNA content of cerebellar and olfactory bulbar neurons of the bat. *Brain Res., 208*, 267-281.

[119] Tollervey, D., Lehtonen, H., Jansen, R., Kern, H. & Hurt, E. C. (1993). Temperature-sensitive mutations demonstrate roles for yeast fibrillarin in pre-rRNA processing, pre-rRNA methylation, and ribosome assembly. *Cell, 72*, 443-457.

[120] Traas, J., Hülskamp, M., Gendreau, E. & Höfte, H. (1998). Endoreplication and development: rule without dividing? *Curr. Opin. Plant Biol., 1*, 498-503.

[121] Urata, Y., Parmelee, S. J., Agard, D. A. & Sedat, J. W. (1995). A three-dimensional structural dissection of *Drosophila* polytene chromosomes. *J. Cell Biol., 131,* 279-295.

[122] Van der Wilt, G. J., Roest, M. & Janse, C. (1987). Neuronal substrates of respiratory behavior and related functions in *Lymnaea stagnalis*. In H. H Boer, W. P. M. Geraerts & J. Joosse J. (Eds.), *Proceedings of the second symposium on molluscan*

neurobiology: Neurobiology, Molluscan models (pp. 292-296). Amsterdam, The Netherlands: North Holland publ. Co.

[123] Varmuza, S., Prideaux, V., Kothary, R. & Rossant, J. (1988). Polytene chromosomes in mouse trophoblast giant cells. *Development, 102,* 127-134.

[124] White, M. J. D. (1973). *Animal Cytology and Evolution.* (Third Edition). Cambridge, UK: Cambridge University Press.

[125] Zimmet, J. & Ravid, K. (2000). Polyploidy: occurrence in nature, mechanisms, and significance for the megakaryocyte-platelet system. *Exp. Hematol., 28*: 3-16.

[126] Zhou, Z. Q. & Hurlin, P. J. (2001). The interplay between Mad and Myc in proliferation and differentiation. *Trends Cell Biol.,* 11, S10-S14.

[127] Zottoli, S. J., Akanki, F. R., Hiza, N. A., Ho-Sang, Jr, D. A., Motta, M., Tan, X. & Watts K. M. (1999). Physiological characterization of supramedullary/dorsal neurons of the cunner *Tautogolabrus adspersus. Biol. Bull., 197*, 239-240.

[128] Zottoli, S. J., Arnolds, D. E. W., Asamoah, N. O., Chenez, C., Fuller, S. N., Hiza, N., A. & Nierman, J. E. (2001). Dye coupling evidence for gap junctions between supramedullary neurons of the cunner *Tautogolabrus adspersus. Biol. Bull., 201*, 277-278.

[129] Zybina, E. V. & Zybina, T. G. (1996). Polytene chromosomes in mammalian cells. *Int. Rev. Cytol., 165*, 53-119.

In: DNA Research Trends
Editor: Marius P. Young, pp. 69-100
ISBN: 978-1-60021-646-6

Chapter III

HUMAN DNA STANDARD REFERENCE MATERIALS DEVELOPED BY THE NATIONAL INSTITUTE OF STANDARDS AND TECHNOLOGY[1,2]

Barbara C. Levin, Kristy L. Richie, Margaret C. Kline, Janette W. Redman and Diane K. Hancock
Biotechnology Division, National Institute of Standards and Technology, Gaithersburg, MD 20899-8311

ABSTRACT

The National Institute of Standards and Technology (NIST), formerly called the National Bureau of Standards (NBS), has developed over 1300 Standard Reference Materials (SRMs). The Office of Standard Reference Materials defines and describes a SRM as a:

"well-characterized material produced in quantity to improve measurement science. It is certified for specific chemical or physical properties, and is issued by NIST with a certificate that reports the results of the characterization and indicates the intended use of the material. A SRM is prepared and used for the following three purposes: 1.To help develop accurate methods of analysis; 2. To calibrate measurement systems used to facilitate exchange of goods, institute quality control, determine performance

[1]This chapter is a contribution of the US National Institute of Standards and Technology (NIST) and is not subject to copyright. Certain commercial equipment, instruments, materials, or companies are identified in this chapter to specify the experimental procedure. Such identification does not imply recommendation or endorsement by NIST, nor does it imply that the materials or equipment identified are the best available for this purpose.

[2] Correspondence should be addressed to: Dr. Barbara C. Levin, 100 Bureau Drive, Stop 8311, National Institute of Standards and Technology, Gaithersburg, Maryland 20899-8311. Telephone: (301) 975-6682, FAX: (301) 330-3447; email: barbara.levin@nist.gov

characteristics, or measure a property at the state-of-the-art limit; and 3. To assure the long-term adequacy and integrity of measurement quality assurance programs" (http://patapsco.nist.gov/srmcatalog/about/ Definitions.htm).

To provide quality assurance in the analysis of human DNA, five Standard Reference Materials - SRM 2390, SRM 2391b, SRM 2392, SRM 2392-I, and SRM 2395 have been developed by NIST and are currently available. Two additional DNA SRMs (# 2394 and # 2399[3]) have been developed and released since the writing of this chapter. Three of these SRMs - the DNA Profiling Standard 2390, the Polymerase Chain Reaction (PCR)-based DNA Profiling Standard 2391b, and the Human Y-Chromosome DNA Profiling Standard 2395 - are intended for use in forensic and paternity identifications, for instructional law enforcement, or non-clinical research purposes. They are not intended for any human/animal clinical diagnostic use. The DNA SRMs, 2392 and 2392-I, are for standardization and quality control when amplifying and sequencing the entire or any segment of human mitochondrial DNA (16,569 base pairs) for forensic identification, disease diagnosis or mutation detection. SRM 2394 is called the Heteroplasmy mtDNA Standard and will permit investigators to determine the limit of their heteroplasmic mutation detection techniques and hopefully, will be useful in promoting the development of new and more sensitive mutation detection methods. Orders and requests for information concerning these SRMs should be directed to: The National Institute of Standards and Technology, Standard Reference Materials Program, 100 Bureau Drive, Mail Stop 2322, Gaithersburg, MD 20899-2322, Telephone: (301) 975-6776, FAX: (301) 948-3730, E-mail: srminfo@nist.gov, Web sites: www.nist.gov/srm, http://ts.nist.gov/ts/htdocs/230/232/232.htm

SRM 2390 – RFLP DNA Profiling Standard

Standard Reference Material 2390 is designed to standardize the methodology used to detect Restriction Fragment Length Polymorphisms (RFLP) using the *Hae* III restriction enzyme. This procedure examines the variable numbers of tandem repeats (VNTR) in highly polymorphic human loci (1,2,3,4). The use of RFLP depends on the presence of a polymorphism which can vary in length. In this case, the *Hae* III restriction cleavage sites flank the tandem repeat sequences of the following commonly probed genetic loci: D1S7, D2S44, D4S139, D5S110, D10S28, and D17S79 and the less commonly probed loci: D7S467, D7Z2, D8S358, D14S13, D17S26, and DYZ1. Different length DNA fragments are generated depending on the number of tandem repeats in an individual. SRM 2390 was originally certified in 1992 and was technically revised in 2001 to update the allelic band sizes based on the latest state-of-the-art forensic practices. While the allelic band sizes are ideally determined by the actual number of tandem repeats, the reported sizes depend on the methodology used to measure them (5). In 1991, the Federal Bureau of Investigation (FBI), the Royal Canadian Mounted Police (RCMP) and many state and local laboratories were using *Hae* III in their RFLP procedures for forensic and paternity testing (6,7). The restriction

---

[3]SRM 2399 provides quality control to clinical laboratories that test human samples for Fragile X and that need to determine the accurate number of CGG trinucleotide repeats.

enzyme *Hae* III was the enzyme of choice because: 1. It is an extremely hardy enzyme, 2. It recognizes a four base sequence (GGCC), 3. It produces relatively small size fragments, and 4. The smaller fragment sizes have better resolution than those obtained with other enzymes (e.g., *Pst* I, *Hinf* I, *Alu* I) (7). In addition, the presence of DNA methylation does not interfere with the enzymatic activity of *Hae* III to recognize or cleave its restriction site.

The DNA Profiling SRM 2390 contains DNA from two sources 1. The cell line, K562 (Life Technologies, Inc., Rockville, MD), which was derived from a female with chronic myelogenous leukemia and 2. TAW (Analytical Genetic Testing Center, Inc., Denver, CO), an apparently healthy male. This SRM provides quality control for the following procedures: 1. Extracting the DNA from the cell pellet, 2. Quantifying the extracted DNA, 3. Determining the restriction enzyme cutting efficiency, 4. Separating the DNA electrophoretically, 5. Blotting the separated DNA onto a nylon membrane using an alkaline transfer solution, 6. Hybridizing the sample DNA to radioactively-labeled DNA probes, 7. Exposing the membrane to X-ray film, and 8. Imaging the developed autoradiogram with a computerized imaging system to determine band sizes. The cells and DNA are provided as 1. A cell pellet containing approximately 3 x 10^6 cells, 2. Extracted genomic DNA and 3. *Hae* III digested DNA. In addition, the SRM contains a DNA molecular size standard for sizing the allele fragments (DNA Analysis Marker System, Gibco BRL Products, Life Technologies, Gaithersburg, MD); a set of quantitative DNA standards consisting of six tubes containing DNA in concentrations from 6 ng to 250 ng; and a visualization marker set based on adenovirus DNA which is cut with restriction enzymes to produce twelve bands ranging from 594 to 35,937 base pairs. The visualization marker is used to assess the DNA separation on the electrophoretic gel. Agarose pretested for compatibility with all the DNA components is also included.

SRM 2390 has 20 components and is packaged in three boxes (containing a total of 19 components) plus a separate container for the agarose. Components #1-5 comprise the DNA Analysis Marker System (Life Technologies, Inc.) for Southern blotting applications using ^{32}P- labeled probes. Components #6-11 plus the loading buffer #19 enable the user to evaluate the quality and quantity of the extracted DNA. Component #18 is an adenovirus DNA visible ladder containing a mixture of digested and undigested adenovirus DNA with fragments ranging from 594 to 35,937 bp. Components 12-17 are the cells and extracted DNA, both *Hae* III treated and untreated from both the female (K562) and male (TAW) sources. These materials and other components of the SRM were obtained from Life Technologies, Inc., Rockville, MD and from Analytical Genetic Testing Center, Inc., Denver, CO.

Interlaboratory Evaluation of SRM 2390

Twenty laboratories participated in an Interlaboratory Evaluation of the 2001 revised SRM 2390 (Table 1) and the data from these laboratories were used to produce the certified allelic band sizes shown in Table 2 and informational band sizes shown in Table 3. Table 2 lists the certified values (expected allelic band sizes and sizing uncertainty) for band size measurements of the cell pellet and extracted K562 and TAW DNA at the six commonly probed genetic loci mentioned above. Table 3 provides informational values for K562 and TAW DNA band sizes at the six less commonly probed loci itemized above. The chromosome location of these various loci as well as those in SRM 2391b are given in Table 4.

Table 1. Participants in the Interlaboratory Recertification of SRM 2390 in 2001

LABORATORY	LOCATION
Arizona Department of Public Safety Crime Laboratory	Phoenix, AZ
Broward County Sheriff's Office, Crime Laboratory	Ft. Lauderdale, FL
Connecticut Forensic Science Laboratory	Meriden, CT
Illinois State Police, Forensic Science Center	Chicago, IL
Illinois State Police, Springfield Forensic Science Lab.	Springfield, IL
Indianapolis-Marion County Forensic Services Agency	Indianapolis, IN
Kentucky State Police Crime Laboratory	Frankfort, KY
LabCorp, Forensic Identity Testing	RTP, NC
Maryland State Police Crime Laboratory	Pikesville, MD
Metro-Dade Police Crime Laboratory	Miami, FL
Minnesota Dept. of Public Safety Bureau of Criminal Apprehension	St. Paul, MN
New York State Police Forensic Investigation Center	Albany, NY
North Carolina State Bureau of Investigation Crime Lab.	Raleigh, NC
Orange County Sheriff's Coroner Dept. DNA Lab.	Santa Ana, CA
Pennsylvania State Police DNA Laboratory	Greensburg, PA
South Carolina Law Enforcement Division DNA Lab.	Columbia, SC
Vermont Forensic Lab., Department of Public Safety	Waterbury, VT
Washington State Patrol Crime Laboratory	Seattle, WA
Washoe County Sheriff's Office Crime Laboratory	Reno, NV
Wisconsin State Crime Laboratory	Milwaukee, WI

All laboratories received the above components, instructions, and a questionnaire. All laboratories used the basic FBI protocol (8) or modifications thereof for RFLP sizing. This protocol consists of the following steps: 1. Extracting DNA from the cell pellet; 2. Quantifying the amount of DNA extracted using the set of quantitative controls supplied with the SRM; 3. Cutting the DNA with a *Hae* III restriction enzyme; 4. Testing the restriction process by gel electrophoresis; 5. Running all electrophoresis on an analytical gel with tris-acetate buffer in the presence of ethidium bromide and appropriate viral DNA markers for assessing proper separation; 6. Blotting the separated DNA onto a nylon membrane using an alkaline transfer solution; 7. Hybridizing the sample DNA to radioactively - labeled DNA probes; 8. Exposing the membrane to X-ray film; and 9. Imaging the developed autoradiogram with a computerized imaging system to determine band sizes. Some of the laboratories participating in the interlaboratory evaluation used probes that were fluorescently labeled rather than radioactively labeled.

Table 2. Certified Values for Band Sizes of SRM 2390 in 2001

DNA Locus [n][a]	DNA Source			
	K562		TAW	
	Band 1	Band 2	Band 1	Band 2
D1S7 [n=20]	4585[b] ± 29.9[c]	4237 ± 23.6	7773 ± 63.2	6886 ± 41.3
D2S44 [n=20]	2905 ± 19.5	1788 ± 14.3	3711 ± 23.2	1288 ± 8.7
D4S139 [n=20]	6474 ± 49.2	3440 ± 22.2	10854 ± 117	8185 ± 64.7
D5S110 [n=17]	3700 ± 23.2	2926 ± 17.4	3343 ± 20.4	1444 ± 11.5
D10S28 [n=19]	1754 ± 11.4	1180 ± 9.7	3935 ± 24.5	1788 ± 10.2
D17S79 [n=15]	1979 ± 14.8	1514 ± 11.1	1753 ± 11.2	1515 ± 11.7

a. Number of data sets received for this locus.
b. Mean number of base pairs calculated from the means of the three preparations (cell, genomic and *Hae* III digested) from all the laboratories.
c. The uncertainty which is the half-width of a 95%/95% statistical tolerance interval for each band. A 95%/95% statistical tolerance interval provides 95% confidence that it contains at least 95% of the relevant population. For each band, this refers to the population of individual results for any of the three preparations, obtained by a population of laboratories similar to those in this interlaboratory study (9).

Table 3. Informational Values for Additional Loci Examined for SRM 2390 in 2001

DNA Locus [n][a]	DNA Source			
	K562		TAW	
	Band 1	Band 2	Band 1	Band 2
D7S467 [n=5]	4677[b]	3217	4496	4339
D7Z2[c] [n=1]	2736		2747	
D8S358 [n=1]	5878	1303	3417	2383
D14S13[d] [n=2]	1642		1579	
D17S26 [n=1]	4823	1358	5514	4852
DYZ1[e] [n=1]			3571	

a. Number of laboratories that submitted data for this specific locus.
b. Mean (in base pairs) of values submitted by all the laboratories that tested these probes.
c. Human-specific monomorphic locus having a known sequence length of 2731 bp.
d. Both K562 and TAW are apparent homozygotes at this locus.
e. Y chromosome locus

Table 4. The chromosome location of the RFLP, STR, and PCR-based markers used in SRM 2390, SRM 2391b, or CODIS.

Chromosome Location	SRM 2390 RFLP Markers	SRM 2391b STR Markers	SRM 2391b CODIS STR Markers	SRM 2391b PCR Markers
1	D1S7	F13B		D1S80
2	D2S44	TPOX D2S1338	TPOX	
3		D3S1358	D3S1358	
4	D4S139	FGA	FGA	
5	D5S110	CSF1PO D5S818	CSF1PO D5S818	
6		F13A1 SE33		DQα
7	D7S467 D7Z2	D7S820	D7S820	
8	D8S358	LPL D8S1179	D8S1179	
9				
10	D10S28			
11		TH01	TH01	
12		vWA	vWA	
13		D13S317	D13S317	
14	D14S13			
15		FES/FPS Penta E		
16		D16S539	D16S539	
17	D17S26			
18		D18S51	D18S51	
19		D19S433		
20				
21		D21S11 Penta D	D21S11	
22				
X				Amelogenin [a]
Y				Amelogenin [a]

[a] Amelogenin is used for gender typing. The X-chromosome gene has 106 base pairs; whereas, the Y-chromosome has 112 base pairs. Therefore, a female (2 X-chromosomes) will have one peak and a male (1 X-chromosome and 1 Y-chromosome) will have two peaks.

In summary, NIST has prepared a standard reference material SRM 2390 to provide quality control and quality assurance to forensic laboratories that are still using RFLP for human identification. Most laboratories are currently using short tandem repeats (STRs) rather than RFLP. The PCR method with STRs is much shorter (days versus weeks with radioactive probes); it requires smaller samples of DNA (1 ng vs. 50-500 ng); the condition of the DNA is less important (may be low molecular weight vs. intact high molecular weight); and finally, the PCR approach may be done on a robot that can process high throughput inhibiting factors in the sample will prevent the amplification and the primers may fail if the primer site contains a mutation. Therefore, some laboratories are still conducting RFLP analysis and can use SRM 2390 in several different ways depending on their quality assurance needs. Components of this SRM are designed to provide assurance that each step of the protocol is functioning properly, but they can also be used for trouble shooting, for calibrating equipment, and for testing the efficacy of new lots of reagents.

SRM 2391b - PCR and STR DNA Profiling Standard

DNA SRM 2391, a PCR-based DNA Profiling Standard, was first issued in 1995 to provide standardization and quality control for forensic identification and paternity testing procedures. As in SRM 2390, it is intended to provide quality assurance for genetic testing, instructional law enforcement, and non-clinical research. It is not intended for use in human or animal clinical diagnostics. SRM 2391 was originally certified for use with the genetic locus D1S80 (pMCT118) (10) and specific primers (11).

The procedures used in the development of SRM 2391 at NIST were based on those of the FBI (PCR-Based Typing Protocols, FBI Laboratory, Quantico, VA, 12/12/94). These procedures consisted of: 1. Extraction of DNA from cells spotted on special filter paper; 2. Amplification of the DNA at the D1S80 locus (11); 3. Separation of the amplified alleles by electrophoresis of the DNA in a gel system that generates good resolution of the allelic ladder; 4. Detection in the gel by silver staining or fluorescent dye followed by image analysis.

The original SRM 2391 contained human cells from two cell culture lines from which DNA could be extracted; genomic DNA from those two cell lines plus eight other individuals; and PCR-amplified DNA from the two cell lines plus four of the eight individuals. A D1S80 allelic ladder for characterization of amplified DNA and a DNA size marker to assure proper electrophoretic separations were also included. SRM 2391 was composed of two boxes (A and B) containing a total of 20 components. The genomic DNA components were from Roche Molecular Systems, Inc., Alameda, CA and the cells, genomic and amplified DNA from cell lines GM 09947A and GM 09948 were obtained from Life Technologies, Inc., Gaithersburg, MD. They were qualified and licensed for forensic use from the National Institute of General Medical Science (NIGMS) repository in Camden, NJ.

Interlaboratory Evaluation of SRM 2391

Prior to certification, various components of SRM 2391 were tested by 20 laboratories (Table 5). Table 6 shows the certified values for the D1S80 alleles determined by three laboratories including NIST. Table 7 shows informational values determined for AmpliType™ HLA-DQα, AmpliType™ PM, TH01, F13A01, vWA, and FES/FPS. AmpliType HLA-DQα was an amplification and typing kit produced by Cetus and was used in the qualitative detection and identification of the human leukocyte antigen (HLA) DQα alleles found on chromosome 6. The kit allowed one to amplify a 239 or 242 bp fragment from the HLA-DQα locus. The typing was performed by hybridization of the amplified DNA to probe strips and permitted visual detection of the probes that were complementary to the unknown DNA sample. AmpliType™ PM PCR Amplification and Typing Kit allowed the typing of 6 polymorphic loci using a reverse dot format. TH01, F13A01, vWA, and FES/FPS are STR loci. See Table 4 for the chromosomal location of these markers.

Table 5. Laboratories Participating in the Interlaboratory Testing for SRM 2391 in 1995

LABORATORY	LOCATION
Alabama Department of Forensic Sciences	Birmingham, AL
Armed Forces DNA Identification Laboratory	Rockville, MD
Broward Sheriff's Crime Laboratory	Fort Lauderdale, FL
California Department of Justice NA Laboratory	Berkeley, CA
Chicago Police Crime laboratory	Chicago, IL
Commonwealth of Virginia Division of Forensic Sciences	Richmond, VA
Connecticut Dept. of Public Safety Forensic Science Lab.	Meriden, CT
Directions des Expertises Judiciaries	Montreal, Quebec, Canada
Dubai Police Headquarters C.I.D. Criminal Laboratory	Dubai, United Arab Emirates
FBI Forensic Science Research & Training Center	Quantico, VA
Forensic Science Associates	Richmond, CA
Georgia Bureau of Investigation Div. of Forensic Sciences	Decatur, GA
Illinois State Police Crime Laboratory	Springfield, IL
Kentucky State Police	Frankfort, KY
Metro-Dade Police Department	Miami, FL
Michigan State Police DNA Laboratory	East Lansing, MI
Minnesota Forensic Science Laboratory	St. Paul, MN
Office of the Chief Medical Examiner	New York, NY
Orange County Sheriff, Coroner Dept. Forensic Science Services	Santa Ana, CA
Roche Biomedical Laboratories, Inc.	Research Triangle Park, NC
Royal Canadian Mounted Police Central Forensic Laboratory	Ottawa, Ontario, Canada
U. CA Berkeley School of Public Health, Forensic Science Group.	Berkeley, CA
U. North Texas, Health Science Center of Fort Worth	Fort Worth, TX

Table 6. Certified Values for SRM 2391[a]

DNA SOURCE	COMPONENT NUMBER	D1S80 ALLELES
Genomic 1	3	28, 31
Genomic 2	4	18, 24
Genomic 3	5	18, 18
Genomic 4	6	21, 24
Genomic 5	7	17, 28
Genomic 6	8	24, 37
Genomic 7	9	24, 28
Genomic 8	10	17, 21
Genomic GM 09947A	11	18, 31
Genomic GM 09948	12	18, 25
Amplified 1	13	28, 31
Amplified 2	14	18, 18
Amplified 3	15	24, 37
Amplified 4	16	17, 21
Amplified GM 09947A	17	18, 31
Amplified GM 09948	18	18, 25
Cells GM 09947A	19	18, 31
Cells GM 09948	20	18, 25

a. The certified values represent the pooled results from analyses performed at two laboratories plus NIST.

RECERTIFICATION OF SRM 2391

Since the issuance of SRM 2391 in 1995, this SRM has been recertified as 2391a in 1998 and as SRM 2391b in 2002. Each ensuing version placed more emphasis on the use of Short Tandem Repeat (STR) loci for human identity rather than D1S80. The initial certificate for SRM 2391 had limited information on only four STR loci, TH01, F13A01, vWA and FES/FPS. The forensic community requested that SRM 2391 include STR information for all the DNA samples and that the STR information include the FBI's 13 CODIS (COmbined DNA Index System) core STR loci (Table 4). The CODIS core loci are CSF1PO, D3S1358, D5S818, D7S820, D8S1179, D13S317, D16S539, D18S51, D21S11, FGA, TH01, TPOX, and vWA. In addition, STR loci F13A01, F13B, FES/FPS, LPL, Penta D, Penta E, D2S1338, D19S433, and SE 33 were certified for SRM 2391b as well as D1S80, AmpliType ® HLA-DQA1, AmpliType ® PM, and Amelogenin. Therefore, a total of 26 loci were analyzed for the recertification of SRM 2391 as SRM 2391b (Tables 4, 8-10).

Table 7. Other Information Values for SRM 2391

DNA Source	Component #	AmpliType HLA-DQα	AmpliType PM (polymarker)	TH01	F13A01	vWA	FES/FPS
Genomic 1	3	2, 4	AA, AB, AA, AA, CC	6, 7	6, 7	N/A	N/A
Genomic 2	4	1.1, 3	AB, BB, AB, AB, AC	8, 9.3	7, 7	N/A	N/A
Genomic 3	5	1.3, 4	BB, AA, BB, BB, AA	9.3, 9.3	3.2, 4	N/A	N/A
Genomic 4	6	4, 4	AB, AA, AB, AA, AB	7, 9	5, 6	N/A	N/A
Genomic 5	7	4, 4	BB, AA, BC, AA, BC	7, 7	5, 7	N/A	N/A
Genomic 6	8	4, 4	AB, AB, AB, AB, AC	9, 9.3	3.2, 5	N/A	N/A
Genomic 7	9	1.2, 4	BB, BB, CC, AB, BB	6, 7	5, 8	N/A	N/A
Genomic 8	10	1.2, 2	BB, BB, AC, AA, BB	7, 8	3.2, 5	N/A	N/A
Genomic GM 09947A	11	4, 4	AA, AB, AB, AA, AC	8, 9.3	6, 16	17, 18	10, 12
Genomic GM 09948	12	1.2, 3	AB, AB, BB, AB, BC	6, 9.3	6, 6	17, 17	11, 11
GM09947A Cells	19	4, 4	AA, AB, AB, AA, AC	8, 9.3	6, 16	17, 18	10, 12
GM09948 Cells	20	1.2, 3	AB, AB, BB, AB, BC	6, 9.3	6, 6	17, 17	11, 11

Table 8. Certified Values for the FBI's CODIS 13 STR Loci

Component No.	Contents	CSF1PO	D3S1358	D5S818	D7S820	D8S1179	D13S317	D16S539	D18S51	D21S11	FGA	TH01	TPOX	vWA
1	Genomic 1	12,12	14,17	12,12	9,10	13,13	11,13	12,14	14,14	29,33.2	21,22	6,7	8,11	17,17
2	Genomic 2	11,12	15,16	12,12	9,10	11,16	8,11	12,12	10,14	29,30	20,22	8,9.3	8,10	14,16
3	Genomic 3	11,12	15,15	11,11	12,13	14,16	11,12	11,12	16,20	28,31.2	23,25	9.3,9.3	8,11	18,19
4	Genomic 4	11,12	15,17	11,11	8,10	14,14	12,12	9,10	18,18	28,30	18,22	7,9	8,9	17,17
5	Genomic 5	10.12	15,18	11,12	8,10	15,16	11,12	9,11	14,16	28,30	23,26	7,7	10,11	16,20
6	Genomic 6	10,13	14,17	12,12	8,11	10,16	12,13	12,13	18,18	28,29	21,26	9,9.3	8,8	16,18
7	Genomic 7	10,11	14,15	11,12	9,9	13,15	11,12	10,10	13,16	28,31.2	23,24	6,7	8,11	16,16
8	Genomic 8	10,12	15,18	12,13	9,10	12,14	9,13	9,11	15,18	30,31	24,28	7,8	8,12	15,17
9	Genomic GM09947A	10,12	14,15	11,11	10,11	13,13	11,11	11,12	15,19	30,30	23,24	8,9.3	8,8	17,18
10	Genomic GM09948	10,11,12	15,17	11,13	11,11	12,13	11,11	11,11	15,18	29,30	24,26	6,9.3	8,9	17,17
11	GM09947A cells	10,12	14,15	11,11	10,11	13,13	11,11	11,12	15,19	30,30	23,24	8,9.3	8,8	17,18
12	GM09948 cells	10,11 or 10,11,12[a]	15,17	11,13	11,11	12,13	11,11	11,11	15,18	29,30	24,26	6,9.3	8,9	17,17

[a]The relative intensity of the 12 allele is less than 7% of that of the dominant 10 allele

Table 9. Certified Values for Additional STR Loci for SRM 2391b

Component	Contents	F13A01	F13B	FES/FPS	LPL	Penta D	Penta E	D2S1338	D19S433	SE33
1	Genomic 1	6,7	10,10	12,12	10,11	10,15	7,12	17,23	13,16.2	20,30.2
2	Genomic 2	7,7	8,10	10,11	10,11	9,11	7,12	17,26	14,16	23.2,28.2
3	Genomic 3	3.2,4	9,10	11,12	11,12	11,12	13,14	20,24	12,14	14.2,26.2
4	Genomic 4	5,6	6,9	10,13	10,12	8,9	5,12	17,23	11,13	22,28.2
5	Genomic 5	5,7	8,9	11,13	10,12	10,13	7,13	17,19	12.2,14	14,30.2
6	Genomic 6	3.2,5	9,10	11,11	10,12	9,12	12,14	25,25	12,14	20,21

Table 9. (Continued)

Component	Contents	F13A01	F13B	FES/FPS	LPL	Penta D	Penta E	D2S1338	D19S433	SE33
7	Genomic 7	5,8	6,8	11,11[a]	11,12	3.2,11	12,16	17,22	13,15.2	13.2,20
8	Genomic 8	3.2,5	6,8	10,11	9,11	8,9	5,10	22,22	12.2,15	16,27.2
9	Genomic GM09947A	6,16	8,10	10,12	11,12	12,12	12,13	19,23	14,15	19,29.2
10	Genomic GM09948	6,6	8,8	11,11	10,12	8,12	11,11	23,23	13,14	23.2,26.2
11	GM09947A cells	6,16	8,10	10,12	11,12	12,12	12,13	19,23	14,15	19,29.2
12	GM09948 cells	6,6	8,8	11,11	10,12	8,12	11,11	23,23	13,14	23.2,26.2

[a] Genomic 7 has a variant allele at the FES/FEP locus. Preliminary sequence analysis indicates sequence changes on either side of the STR.

Table 10. Certified Values for AmpliType® HLA-DQA1, AmpliType® PM, and Amelogenin for SRM 2391b

DNA Source	Component Number	AmpliType® HLA-DQA1	AmpliType® PM					Amelogenin
Genomic 1	1	2,4.1	AA	AB	AA	AA	CC	XY
Genomic 2	2	1.1,3	AB	BB	AB	AB	AC	XX
Genomic 3	3	1.3,4.1	BB	AA	BB	BB	AA	XY
Genomic 4	4	4.1,4.2/4.3	AB	AA	AB	AA	AB	XX
Genomic 5	5	4.1,4.1	BB	AA	BC	AA	BC	XX
Genomic 6	6	4.1,4.1	AB	AB	AB	AB	AC	XX
Genomic 7	7	1.2,4.1	BB	BB	CC	AB	BB	XY
Genomic 8	8	1.2,2	BB	BB	AC	AA	BB	XX
Genomic GM09947A	9	4.1,4.2/4.3	AA	AB	AB	AA	AC	XX
Genomic GM09948	10	1.2,3	AB	AB	BB	AB	BC	XY
GM09947A Cells	11	4.1,4.2/4.3	AA	AB	AB	AA	AC	XX
GM09948 Cells	12	1.2,3	AB	AB	BB	AB	BC	XY

For certified values for D1S80, see table 6.

There are a number of advantages using PCR methods with STR loci over RFLP as mentioned above. The more STR loci that one examines, the greater the power of discrimination and much work has been done on multiplexing the STR loci, so multiple loci can be examined at the same time.

The laboratories that participated in an interlaboratory evaluation of SRM 2391b were: NIST, Gaithersburg, MD; Pennsylvania State Police DNA Laboratory, Greensburg, PA; Oregon State Police Forensic Laboratory, Portland, OR; Promega Corp., Madison, WI; and Applied Biosystems, Foster City, CA. The values in Tables 8-10 are the pooled results from all the laboratories that participated in the interlaboratory evaluation. The amplification kits that were used at NIST to genotype the various STR loci are shown in Table 11.

Table 11. Amplification Kits used at NIST to Genotype STR Loci for SRM 2391b

STR Locus	Applied Biosystems				Promega			
	Identi-filer	Profiler Plus	COfiler	SGM Plus	Power Plex 16	Power Plex ES	Power Plex 16 Bio	FFFL
Amelogenin	X	X	X	X	X	X	X	
CSF1PO	X		X		X		X	
D2S1338	X			X				
D3S1358	X		X	X	X	X	X	
D5S818	X	X			X		X	
D7S820	X	X	X		X		X	
D8S1179	X	X		X	X	X	X	
D13S317	X	X	X		X		X	
D16S539	X			X	X		X	
D18S51	X	X		X	X	X	X	
D19S433	X			X				
D21S11	X	X		X	X	X	X	
F13A01								X
F13B								X
FES/FPS								X
FGA	X	X		X	X	X	X	
LPL								X
PENTA D					X		X	
PENTA E					X		X	
SE33						X		
TH01	X		X	X	X	X	X	
TPOX	X		X		X		X	
vWA	X	X		X	X	X	X	

Twelve components are included in each SRM 2391b unit. Components #1 through #10 each contain 20 µL of genomic DNA at a concentration of approximately 1 ng/µL. Vials #1 through 8 contain Genomic DNA 1 through Genomic DNA 8, respectively. Vial #9 contains

Genomic DNA from GM09947A and Vial #10 contains Genomic DNA from GM09948. Components #11 and #12 consist of GM09947A cells and GM09948 cells at a concentration of 2 x 10^5 on a 7 mm Scleicher & Schull 903 ™ filter paper circle 1, respectively.

In summary, SRM 2391b is designed to provide quality assurance to laboratories that each step of the PCR methodology to examine STR loci is operating correctly and within the proper limits. The examination of multiple STR loci increases the power of discrimination between individuals. The recent improvement in the ability to multiplex and examine many STR loci simultaneously also enhances the utility of this method.

SRM 2392 and SRM 2392-I - Human Mitochondrial Amplification and Sequencing Standards

Human mitochondrial DNA (mtDNA) was completely sequenced in 1981 and found to consist of a circular double-stranded molecule containing 16,569 base pairs (12). This sequence determined by Anderson *et al.* is referred to as the Cambridge Reference Sequence (CRS) and was a consensus sequence based primarily on the results from a human placenta, partly on the sequence from HeLa cells, and in five nucleotide positions where the human results were ambiguous, the authors inserted the results from the bovine mtDNA that they had also sequenced. In 1999, Andrews *et al.* (13) rediscovered and resequenced the original placenta and found that the 1981 sequence contained a number of rare polymorphisms that were specific for that placenta plus a number of errors.

Each human cell can have a few dozen to several thousand molecules of mtDNA (14, 15). Sequence analysis of mtDNA is being used by the forensic community for human identification especially in those cases where genomic DNA is highly degraded or non-existent (16, 17). Forensic analysis to distinguish between individuals is primarily based on the considerable sequence variation found in the two hypervariable regions (HV1, HV2) located in the non-coding displacement loop (D-loop) or control region. These are the areas designated by primer sets 1, 57 and 58 (Table 12). The medical community is also using sequence analysis of mtDNA for diagnoses of diseases associated with specific mutations and deletions (18). A third area of research which is largely unexplored and which needs sequence analysis is the examination of the mutagenic effects of chemical and physical agents on mtDNA (19, 20).

Human mtDNA SRMs 2392 and 2392-I were developed at NIST to be used for quality control when amplifying and sequencing DNA for forensic identifications, medical diagnostics, or mutation detection. SRM 2392, which became available in 1999, compares the entire mtDNA sequences of two lymphoblastoid cell lines from two apparently normal individuals (CHR and 9947A[4]) (21) with the 1981 Cambridge Reference Sequence. SRM 2392 also contains the cloned DNA from the CHR HV1 region which contains a C-stretch, an area beyond which sequencing becomes very difficult. In 2003, we developed human mtDNA amplification and sequencing SRM 2392-I (22,23), which contains the DNA from cell line HL-60[5] (a promyelocytic cell line prepared from the peripheral blood leukocytes of a female with acute promyelocytic leukemia). The sequencing results from SRM 2392-I are compared

---

[4] Life Technologies, Inc., Gaithersburg, MD. This cell line is also used in SRM 2391b.

[5] DNA was prepared by the Professional Services Department of American Type Culture Collection (ATCC, Manassas, VA).

with both the 1981 and 1999 Cambridge Reference Sequences and the sequences obtained for SRM 2392. The information on the entire mtDNA sequences (but not the DNA) from GM03798[6] (a normal human lymphoblast cell line) and GM10742A[6] [a cell line from an individual with Leber Hereditary Optic Neuropathy (LHON)] are provided for information and comparison purposes. Both human mtDNA SRMs 2392 and 2392-I complement one another and provide additional DNA templates for quality assurance when amplifying and sequencing unknown specimens.

Fifty-eight unique primer sets were designed to allow any area or the entire mtDNA (16,569 base pairs) to be amplified and sequenced using the same conditions. The sequences of these primers shown in Table 12 are the same as those used for SRM 2392 (21) with the exception of the reverse primer of set 51. The original reverse primer of set 51 was designed to bind to the CRS sequence. Andrews *et al.* (13) in their reevaluation of the placenta originally used to sequence human mtDNA in 1981 (12) found a C, not a G, at np 14368. Therefore, the reverse primer was redesigned.

With one exception, none of the differences in the coding region of the three normal templates (CHR, 9947A, GM03798) correspond to published mutations associated with specific diseases; although some of these differences did result in amino acid changes when compared with the Cambridge Reference Sequence (12, 13). However, HL-60 did have four mutations that corresponded with mutations that have been associated with the disease Leber Hereditary Optic Neuropathy (LHON), which causes blindness in young adults (18). Due to this finding, we also sequenced the entire mtDNA of a cell line (GM10742A) derived from a person diagnosed with LHON to determine how many mutations these two cell lines would have in common. Two of the four LHON associated mutations were in both cell lines. GM10742A had an additional two LHON mutations. All the differences from the CRS (both 1981 and 1999) from all five templates that were sequenced are shown in Fig. 1 and Table 13. This table also indicates if the mutation produces an amino acid change in the resultant protein. Since we have no information on the identity of HL-60, we do not know if that individual also developed LHON.

SRM 2392 was revalidated at NIST in 2003 and the expiration date was extended to 2008.

INTERLABORATORY EVALUATION OF SRM 2392 AND 2392-I

An interlaboratory evaluation of the amplification, sequencing, and data analysis of the CHR template for SRM 2392 was conducted by The Bode Technology Group, Inc., Sterling, VA; IIT Research Institute, Newington, VA; and Lark Technologies, Inc., Houston, TX plus NIST. Another interlaboratory evaluation of the amplification, sequencing, and data analysis of HL-60 for SRM 2392-I was conducted by the FBI Laboratory, FBI Academy, Quantico, VA; Armed Forces DNA Identification Laboratory (AFDIL), Armed Forces Institute of

[6] NIGMS Human Genetic Mutant Cell Repository, Coriell Institute for Medical Research, 401 Haddon Ave., Camden, NJ.

Pathology, Rockville, MD; and the Georgia Bureau of Investigation (GBI), Decatur, GA plus NIST. Corroboration of the SRM results provides quality assurance that any unknown mtDNA is also being amplified and sequenced correctly.

For SRM 2392, each participant of the interlaboratory evaluation was provided:

1. Two tubes of DNA from the CHR cell culture line, one contained extracted DNA ready for PCR amplification of the entire mtDNA, and the other contained the cloned DNA ready for cycle sequencing of the HV1 region (this DNA did *not* need to be PCR amplified).
2. Fifty-eight sets of primers labeled with either F# (forward primer) or R# (reverse primer). Forward and reverse primers with the same number were paired and numbered from the 5' end. Primers were diluted to 10 µmol/L and ready for use. Also sent was the -21M13 primer to do the sequencing of the cloned HV1 region of the CHR DNA.
3. A copy of the protocol used at NIST to amplify and sequence the DNA. The participants in the interlaboratory evaluation, however, were free to use any protocol with which they were familiar and
4. A form table to record the results. This table provided the number of the Primer Set, the region that each Primer Set amplified, and the length of the amplified region. The laboratory was asked to fill in the differences found when they compared the sequence that they determined for the CHR template with that of the Cambridge Reference Sequence (12).

All of the laboratories examining CHR essentially followed the NIST protocol. Any changes to the protocol are listed in the section on Materials and Methods in reference 21. Each laboratory was instructed to amplify and sequence the 58 areas designated by the 58 primer sets and also to sequence the cloned DNA for the HV1 region. Labs 1, 2 and 3 found essentially the same polymorphisms. Laboratory 4 had less experience with sequencing mtDNA and did find differences that the other laboratories did not observe. This interlaboratory evaluation was considered successful in that most of the laboratories found the same results. The differences noted by Laboratory 4 confirm and emphasize the need for a standard reference material for sequencing mtDNA. If Laboratory 4 had the NIST mtDNA SRM 2392 and had run it simultaneously with their unknown sample, they would have realized that they were finding an undue number of differences and could have reexamined their procedures to try to determine the reason for these differences.

For SRM 2392-I, each participant in the interlaboratory evaluation was asked to amplify and sequence the entire mtDNA of HL-60. NIST provided: 1. A tube of DNA containing the extracted DNA ready for PCR amplification, 2. The 58 sets of primers labeled with either F# (forward primer) or R# (reverse primer), 3. A table to record the results, and 4. Any other supplies that were needed and requested by the participants. They were allowed to use any protocol for amplification or sequencing that they wished, but were requested to provide a copy of that protocol to NIST. NIST also requested copies of the electropherograms to enable us to resolve any discrepancies; although, as it turned out, all four laboratories obtained identical results. The differences in the methodologies used by the laboratories participating in the interlaboratory evaluation of HL-60 are described in references 22 and 23.

Table 12. Primer Sets Used for the Amplification and Sequencing of mtDNA in SRMs 2392 and 2392-I

Primer Set Number	Nucleotide Positions of Forward and Reverse Primers	Primer Sequence
1	F15	CACCCTATTAACCACTCACG
(HV2)	R484	TGAGATTAGTAGTATGGGAG
2	F361	ACAAAGAACCCTAACACCAGC
	R921	ACTTGGGTTAATCGTGTGACC
3	F756	CATCAAGCACGCAGCAATG
	R1425	AATCCACCTTCGACCCTTAAG
4	F873	GGTTGGTCAATTTCGTGCCAG
	R1425	AATCCACCTTCGACCCTTAAG
5	F1234	CTCACCACCTCTTGCTCAGC
	R1769	GCCAGGTTTCAATTTCTATCG
6	F1587	TGCACTTGGACGAACCAGAG
	R2216	TGTTGAGCTTGAACGCTTTC
7	F1657	CTTGACCGCTCTGAGCTAAAC
	R2216	TGTTGAGCTTGAACGCTTTC
8	F1993	AAACCTACCGAGCCTGGTG
	R2216	TGTTGAGCTTGAACGCTTTC
9	F2105	GAGGAACAGCTCTTTGGACAC
	R2660	AGAGACAGCTGAACCCTCGTG
10	F2417	CACTGTCAACCCAACACAGG
	R3006	ATGTCCTGATCCAACATCGAG
11	F2834	CCCAACCTCCGAGCAGTACATG
	R3557	AGAAGAGCGATGGTGAGAGC
12	F2972	ATAGGGTTTACGACCTCGATG
	R3557	AGAAGAGCGATGGTGAGAGC
13	F3234	AGATGGCAGAGCCCGGTAATC
	R3557	AGAAGAGCGATGGTGAGAGC
14	F3441	ACTACAACCCTTCGCTGACG
	R3940	TGAAGCCTGAGACTAGTTCGG
15	F3635	GCCTAGCCGTTTACTCAATCC
	R4162	TGAGTTGGTCGTAGCGGAATC
16	F3931	TCAGGCTTCAACATCGAATACG
	R4728	TTATGGTTCATTGTCCGGAGAG
17	F4183	TTTCTACCACTCACCCTAGCATTAC
	R4728	TTATGGTTCATTGTCCGGAGAG
18	F4392	CCCATCCTAAAGTAAGGTCAGC
	R4983	GGTTTAATCCACCTCAACTGCC
19	F4447	TTGGTTATACCCTTCCCGTAC
	R4982	GTTTAATCCACCTCAACTGCC
20	F4797	CCCTTTCACTTCTGAGTCCCAG
	R5553	AGGGCTTTGAAGGCTCTTG
21	F4976	ATTAAACCAGACCCAGCTACG
	R5553	AGGGCTTTGAAGGCTCTTG
22	F5318	CACCATCACCCTCCTTAACC
	R5882	GCTGAGTGAAGCATTGGACTG
23	F5700	TAAGCACCCTAATCAACTGGC
	R6262	GCCTCCACTATAGCAGATGCG

Table 12. (Continued)

Primer Set Number	Nucleotide Positions of Forward and Reverse Primers	Primer Sequence
24	F5999	TCTAAGCCTCCTTATTCGAGC
	R6526	ATAGTGATGCCAGCAGCTAGG
25	F6242	CGCATCTGCTATAGTGGAGG
	R6526	ATAGTGATGCCAGCAGCTAGG
26	F6426	GCCATAACCCAATACCAAACG
	R7030	TGGGCTACAACGTAGTACGTG
27	F6744	GGCTTCCTAGGGTTTATCGTG
	R7255	TTTCATGTGGTGTATGCATCG
28	F7075	GAGGCTTCATTCACTGATTTCC
	R7792	GGGCAGGATAGTTCAGACGG
29	F7215	CGACGTTACTCGGACTACCC
	R7792	GGGCAGGATAGTTCAGACGG
30	F7645	TATCACCTTTCATGATCACGC
	R8215	GACGATGGGCATGAAACTG
31	F7901	TGAACCTACGAGTACACCGACTAC
	R8311	AAGTTAGCTTTACAGTGGGCTCTAG
32	F8164	CGGTCAATGCTCTGAAATCTGTG
	R8669	CATTGTTGGGTGGTGATTAGTCG
33	F8539	CTGTTCGCTTCATTCATTGCC
	R9059	GTGGCGCTTCCAATTAGGTG
34	F8903	CCCACTTCTTACCACAAGGC
	R9403	GTGCTTTCTCGTGTTACATCG
35	F9309	TTTCACTTCCACTCCATAACGC
	R9848	GAAAGTTGAGCCAATAATGACG
36	F9449	CGGGATAATCCTATTTATTACCTCAG
	R9995	AGAGTAAGACCCTCATCAATAGATGG
37	F9754	AGTCTCCCTTCACCATTTCCG
	R10275	AAAGGAGGGCAATTTCTAGATC
38	F10127	ACTACCACAACTCAACGGCTAC
	R10556	GGAGGATATGAGGTGTGAGCG
39	F10386	GGATTAGACTGAACCGAATTGG
	R11166	CATCGGGTGATGATAGCCAAG
40	F10704	GTCTCAATCTCCAACACATATGG
	R11267	TGTTGTGAGTGTAAATTAGTGCG
41	F11001	AACGCCACTTATCCAGTGAACC
	R11600	CTGTTTGTCGTAGGCAGATGG
42	F11403	GACTCCCTAAAGCCCATGTCG
	R11927	TTGATCAGGAGAACGTGGTTAC
43	F11760	ACGAACGCACTCACAGTCG
	R12189	AAGCCTCTGTTGTCAGATTCAC
44	F11901	TGCTAGTAACCACGTTCTGGTG
	R12876	GATATCGCCGATACGGTTG
45	F12357	AACCACCCTAACCCTGACTTCC
	R12876	GATATCGCCGATACGGTTG
46	F12601	TTCATCCCTGTAGCATTGTTCG
	R13123	AGCGGATGAGTAAGAAGATTCC
47	F12793	TTGCTCATCAGTTGATGATACG
	R13343	TTGAAGAAGGCGTGGGTACAG
48	F13188	CACTCTGTTCGCAGCAGTATG
	R13611	TCGAGTGCTATAGGCGCTTGTC

Table 12. (Continued)

Primer Set Number	Nucleotide Positions of Forward and Reverse Primers	Primer Sequence
49	F13518	CATCATCGAAACCGCAAAC
	R13935	TGTGATGCTAGGGTAGAATCCG
50	F13715	GAAGCCTATTCGCAGGATTTC
	R14118	TGGGAAGAAGAAAGAGAGGAAG
51[a]	F13899	TTTCTCCAACATACTCGGATTC
	R14388	TTAGCGATGGAGGTAGGATTCG (old primer)
	R14388	TTAGCGATGGAGGTAGGATT**G**G (new primer)
52	F14189	ACAAACAATGGTCAACCAGTAAC
	R14926	TGAGGCGTCTGGTGAGTAGTGC
53	F14470	TCCAAAGACAACCATCATTCC
	R14996	CGTGAAGGTAGCGGATGATTC
54	F14909	TACTCACCAGACGCCTCAACCG
	R15396	TTATCGGAATGGGAGGTGATTC
55	F15260	AGTCCCACCCTCACACGATTC
	R15774	ACTGGTTGTCCTCCGATTCAGG
56	F15574	CGCCTACACAATTCTCCGATC
	R16084	CGGTTGTTGATGGGTGAGTC
57 (HV1)	F15971	TTAACTCCACCATTAGCACC
	R16451	GCGAGGAGAGTAGCACTCTTG
58	F16097	TACATTACTGCCAGCCACCATG
	R336	TTAAGTGCTGTGGCCAGAAG
-21M13	F	TGTAAAACGACGGCCAGT

[a] During the development of SRM 2392-I, reverse primer 51 was changed. When sequencing this area of human mtDNA, one should use the new reverse primer 51.

Differences Between SRM 2392 and SRM 2392-I Templates, GM03798, GM10742A, and the Cambridge Reference Sequence

The Cambridge Reference Sequence (CRS) for human mtDNA was first published in 1981 and revised in 1999 (12, 13). All investigators, who subsequently examined human mtDNA, have used the numbering system of the CRS and have compared their sequence findings to that sequence. The CRS DNA, however, is not available for use as a positive control during actual experiments; whereas, NIST SRM 2392 and 2392-I are available. Table 13 shows the mtDNA differences compared to the CRS that were found at NIST with all five templates - CHR, 9947A, GM03798, HL-60 and GM10742A. Compared to the 1981 CRS, CHR mtDNA had 13 differences in the non-coding regions and 33 differences in the coding regions, 9947A mtDNA had 9 differences in the non-coding regions and 23 differences in the coding regions, HL-60 had 11 differences in the non-coding regions and 33 differences in the coding regions. GM03798 and GM 10742A, whose data are included for comparison and information, had 4 and 12 differences in the non-coding regions, respectively and 19 and 14 differences in the coding regions, respectively. All of these differences are shown in Figure 1 along with many of the LHON disease mutation sites that have been noted in the literature (18). With one exception, none of the base pair changes found in the coding regions of the three normal templates sequenced at NIST correlate with any of the changes found associated with LHON or other published disease states

Table 13. Certified Human mtDNA Sequence Differences from the Cambridge Reference Sequence (CRS) Found in the Two Templates (CHR and 9947A) in NIST SRM 2392, One Template (HL-60) in NIST SRM 2392-I and in GM03798 and GM10742A.

Comparison with the Cambridge Reference Sequence (CRS)								
CRS		Template	Template	Template	Template	Template	Amino acid change	Region
# [a]	Base[b] 1981/1999	CHR[d]	9947A[d]	HL-60	GM03798	GM10742A		
73	A	G	-	G	-	G		HV2
93	A	-	G	-	-	-		HV2
150	C	-	-	T	-	-		HV2
152	T	-	-	C	-	-		HV2
185	G	-	-	-	-	A		HV2
195	T	C	C	-	-	-		HV2
204	T	C	-	-	-	-		HV2
207	G	A	-	-	-	-		HV2
214	A	-	G	-	-	-		HV2
228	G	-	-	-	-	A		HV2
263*R	A	G	G	G	G	G		HV2
295	C	-	-	T	-	T		HV2
303-309	-	C (ins)	CC (ins)	-	-	-		HV2
311-315*R	-	C (ins)	C (ins)	C (ins)	C (ins)	C (ins)		HV2
462	C	-	-	-	-	T		HV2
482	T	-	-	-	-	C		HV2
489	T	-	-	C	-	C		HV2
709	G	A	-	-	A	-		12sRNA
750 *R	A	G	G	G	G	G		12sRNA
1438*R	A	G	G	G	G	G		12sRNA
1719	G	A	-	-	-	-		16sRNA
2706	A	G	-	G	-	G		16sRNA
2841	T	-	-	-	-	A		16sRNA
3010	G	-	-	-	A	A		16sRNA

Table 13. (Continued)

Comparison with the Cambridge Reference Sequence (CRS)								
CRS		Template CHR[d]	Template 9947A[d]	Template HL-60	Template GM03798	Template GM10742A	Amino acid change	Region
# [a]	Base[b] 1981/1999							
3106-3107*E	C/del	del C	del C	del C	del C	del C		16sRNA
3394	T	-	-	-	-	C	Tyr → His	ND1 LHON
3423*E	G/T	T	T	T	T	T	Silent	ND1
4135	T	-	C	-	-	-	Tyr → His	ND1
4216	T	-	-	C	-	C	Tyr → His	ND1 LHON
4769*R	A	G	G	G	G	G	Silent	ND2
4985*E	G/A	A	A	A	A	A	Silent	ND2
5186	A	G	-	-	-	-	Silent	ND2
5228	C	-	-	G	-	-	Silent	ND2
5633	C	-	-	T	-	-		tRNA Ala
6221	T	C	-	-	-	-	Silent	COI
6371	C	T	-	-	-	-	Silent	COI
6791	A	G	-	-	-	-	Silent	COI
6849[h]	A	G(0.3A)[h]	-	-	-	-	Thr → Ala[h]	COI
7028	C	T	-	T	-	T	Silent	COI
7476	C	-	-	T	-	-		tRNA Ser
7645	T	-	C	-	-	-	Silent	COII
7861	T	-	C	-	-	-	Silent	COII
8020	G	-	-	-	-	A	Silent	COII
8448	T	-	C	-	-	-	Met → Thr	ATPase 8
8503	T	C	-	-	-	-	Silent	ATPase 8
8860*R	A	G	G	G	G	nd	Thr → Ala	ATPase6
9315	T	-	C	-	-	-	Phe → Leu	COIII
9559*E	G/C	C	C	C	C	C	Arg → Pro	COIII
10172	G	-	-	A	-	-	Silent	ND3
10398	A	-	-	G	-	G	Thr → Ala	ND3
11251	A	-	-	G	-	G	Silent	ND4
11287	T	-	-	-	-	C	Silent	ND4

Table 13. (Continued)

Comparison with the Cambridge Reference Sequence (CRS)								
CRS		Template CHR[d]	Template 9947A[d]	Template HL-60	Template GM03798	Template GM10742A	Amino acid change	Region
# [a]	Base[b] 1981/1999							
11335*E	T/C	C	C	C	C	C	Silent	ND4
11719	G	A	-	A	-	A	Silent	ND4
11778	G	-	-	-	-	A	Arg → His	ND4 LHON
11878	T	C	-	-	-	-	Silent	ND4
12071[het]	T	-	-	C/T[het]	-	-	Phe→Leu[het]	ND4
12612	A	G	-	G	-	G	Silent	ND5
12705	C	T	-	-	-	-	Silent	ND5
13572	T	-	C	-	-	-	Silent	ND5
13702*E	G/C	C	C	C	C	C	Gly → Arg	ND5
13708	G	A	-	A	-	A	Ala → Thr	ND5 LHON
13759	G	-	A	-	-	-	Ala → Thr	ND5
13966	A	G	-	-	-	-	Thr → Ala	ND5
14199*E	G/T	T	T	T	T	T	Pro → Thr	ND6
14272*E	G/C	C	C	C	C	C	Phe → Leu	ND6
14365*E	G/C	C	C	C	C	C	Silent	ND6
14368*E	G/C	C	C	C	C	C	Phe → Leu	ND6
14470	T	C	-	-	-	-	Silent	ND6
14569	G	-	-	A	-	-	Silent	ND6
14766*E	T/C	T	C	T	C	T	Ile → Thr	ND6
14798	T	-	-	-	-	C	Phe → Leu	CYT B
15257	G	-	-	A	-	-	Asp→ Asn	CYT B LHON
15326*R	A	G	G	G	G	G	Thr → Ala	CYT B
15452	C	-	-	A	-	A	Leu → Ile	CYT B
15646	C	-	-	-	T	-	Silent	CYT B
15812	G	-	-	A	-	-	Val → Met	CYT B LHON
16069	C	-	-	T	-	T		HV1
16126	T	-	-	-	-	C		HV1

Table 13. (Continued)

Comparison with the Cambridge Reference Sequence (CRS)								
CRS		Template CHR[d]	Template 9947A[d]	Template HL-60	Template GM03798	Template GM10742A	Amino acid change	Region
# [a]	Base[b] 1981/1999							
16183	A	C	-	-	-	-		HV1
16184-93	-	C (ins)	-	-	-	-		HV1
16189	T	C	-	-	-	-		HV1
16193	C	-	-	T	-	-		HV1
16223	C	T	-	-	-	-		HV1
16278	C	T	-	T	-	-		HV1
16292	C	-	-	-	-	T		HV1
16311	T	-	C	-	-	-		HV1
16357	T	-	-	-	C	-		
16362	T	-	-	C	-	-		HV1
16519	T	C	C	-	C	nd		HV1

a : Numbers correspond to Cambridge Reference Sequence (12).
b: Base found in 1981 (12)/ Base found in 1999 (13).
d: SRM 2392 (21, 23).
e: SRM 2392-I (22, 23).
- : Base pair same as in 1981 Cambridge Reference Sequence (12).
h : Possible heteroplasmic site. This heteroplasmy seen in the mtDNA from the first CHR cell culture line is not seen in the mtDNA from the second CHR cell culture line. The second CHR cell culture line agrees with the CRS at np 6849. It is DNA from the second CHR cell culture line that is supplied in NIST SRM 2392.
*R: Rare polymorphisms in Cambridge Reference Sequence discovered by reanalysis of original placenta (13).
*E: Error in Cambridge Reference Sequence discovered by reanalysis of original placenta (13).
del: Deletion
ins: Insertion
het: Heteroplasmy found in HL-60 at np 12071.
HV1: Non-coding region found from 16024 and 16569.
HV2: Non-coding region found from 1 and 576.

CHR DNA: Sequence based on two amplifications and cycle sequencing procedures with DNA from the first cell culture line and at least one amplification and cycle sequencing procedure with DNA from the second cell culture line.
9947A DNA: Sequence based on two amplifications and cycle sequencing procedures.
HL-60 DNA: Sequence based on two amplifications and cycle sequencing procedures in both the forward and reverse directions for a total of 4 sequences
ATPase 6: ATP synthase 6
ATPase 8: ATP synthase 8
CYTB: Cytochrome B
COI: Cytochrome C Oxidase I
COII: Cytochrome C Oxidase II
COIII: Cytochrome C Oxidase III
ND1: NADH dehydrogenase 1
ND2: NADH dehydrogenase 2
ND3: NADH dehydrogenase 3
ND4: NADH dehydrogenase 4
ND5: NADH dehydrogenase 5
ND6: NADH dehydrogenase 6

In summary, the mtDNA genome of HL-60, a promyelocytic leukemia cell line, has been completely sequenced by four laboratories and is now available as NIST SRM 2392-I to the forensic and medical communities. NIST human mtDNA SRM 2392 also continues to be available and includes the DNA from two apparently healthy individuals. Both SRM 2392 and 2392-I contain all the information (e.g., the sequences of 58 unique primer sets) needed to use these SRMs as positive controls for the amplification and sequencing of any DNA. Compared to the templates in SRM 2392, the HL-60 mtDNA in SRM 2392-I has two tRNA differences and more polymorphisms resulting in amino acid changes. Four of these HL-60 mtDNA polymorphisms have been associated with Leber Hereditary Optic Neuropathy (LHON), one as an intermediate mutation and three as secondary mutations. The mtDNA from cell line GM10742A from an individual with LHON was also completely sequenced for comparison and contained some of the same LHON mutations. The combination of these particular LHON associated mutations is also found in phylogenetic haplogroup J and its subset, J_2, and may only be indicative that HL-60 belongs to haplogroup J, one of nine haplogroups that characterize Caucasian individuals of European descent or may mean that haplogroup J is more prone to LHON. Both these mtDNA SRMs will provide enhanced quality control in forensic identification, medical diagnosis, and single nucleotide polymorphism (SNP) detection to the scientific and medical communities when they sequence human mtDNA.

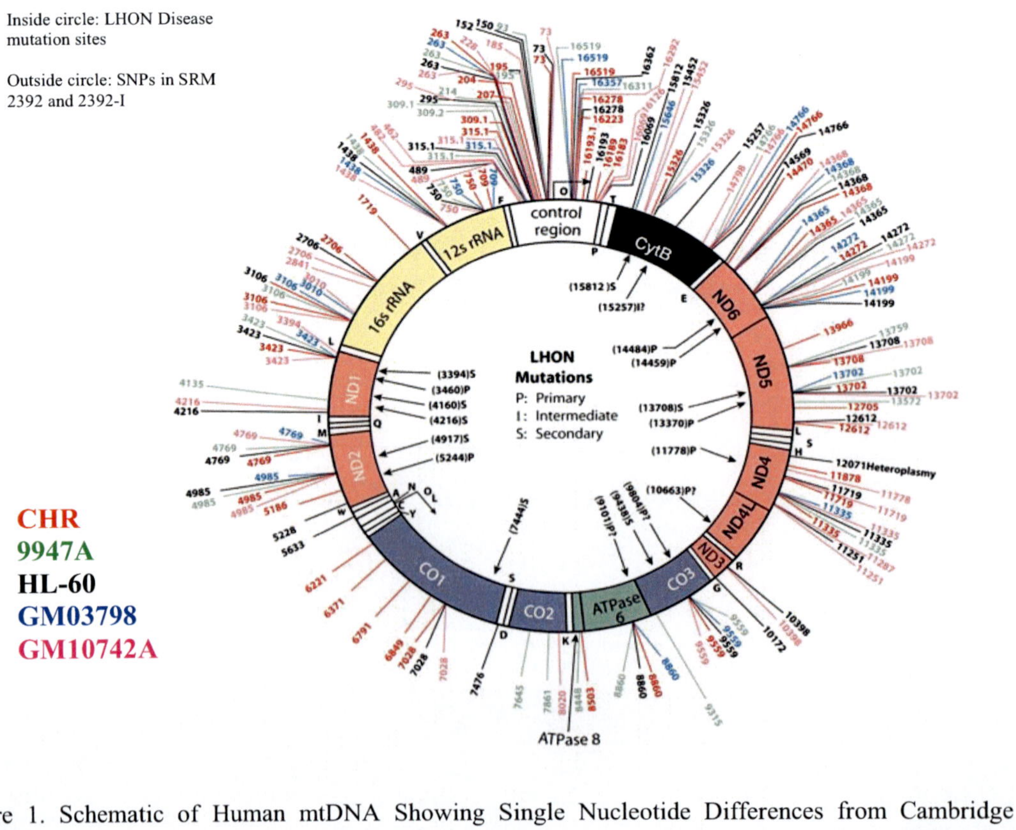

Figure 1. Schematic of Human mtDNA Showing Single Nucleotide Differences from Cambridge Reference Sequence found in SRM 2392, SRM 2392-I, and DNA templates from cell lines GM03798 and GM10742A.

SRM 2395 – Human Y-Chromosome DNA Profiling Standard

SRM 2395 is the Human Y-Chromosome DNA Profiling Standard and is designed to provide quality control to scientists conducting genetic tests involving the polymerase chain reaction (PCR), to law enforcement agents using it as an instructional aid, and to researchers engaging in non-clinical studies that involve the human Y-chromosome. It is not intended for any human or animal clinical diagnostic use. The Y-chromosome is a paternally-transmitted chromosome inherited by males only and is the factor that determines the gender of an individual's offspring. Females have two X chromosomes (one from each parent), while males have one X (from their mother) and one Y (from their father). Therefore, Y-chromosome short tandem repeat (STR) loci are useful in forensic cases such as sexual assaults to distinguish between the DNA of the male assailant and the female victim. These STRs can also be used in paternity testing, identity verification, evolutionary studies to determine population origins, and male lineage genealogy studies (e.g., to determine if the descendants of Sally Hemmings also descended from Thomas Jefferson or other males from the same paternal lineage). Thus, two additional uses of Y-chromosome SRM 2395 are to serve as positive controls in paternity determinations and in identifications of male perpetrators in rape cases. By obtaining the certified results with SRM 2395, the investigators will have the quality assurance that their results with their unknown samples are correct.

SRM 2395 is composed of well-characterized human genomic DNA from five males and one female (negative control) and consists of the following specific genetic STR loci that have been certified by sequencing of the various alleles: DYS19, DYS385 a/b, DYS388, DYS389 I/II, DYS390, DYS391, DYS392, DYS393, DYS426, DYS435, DYS436, DYS437, DYS438, DYS439, DYS447, DYS448, DYS460, DYS461, DYS462, and Y-GATA-H4 (Tables 14, 15) (24 – 35). Table 16 provides informational values for five additional Y STR loci (nine allele calls) that have been typed, but not yet sequenced: DYS450, DYS456, DYS458 DYS464 a/b/c/d, and YCAII a/b. Additional information on each Y chromosome marker can be found at a NIST-sponsored database on the Internet at http://www.cstl.nist.gov/biotech/strbase and the positions of these STR markers along the Y chromosome are shown in Figure 2. Table 17 contains a summary of the sequence information for the 22 sequenced Y STR loci.

For this SRM, genomic DNA components were extracted from whole blood obtained from Millennium Biotech, Inc., Ft. Lauderdale, FL. NIST used four commercial kits to obtain typing results: The ReliaGene Y-PLEX ™ 6 kit (30) was used to type DYS19, DYS389II, DYS390, DYS391, DYS393, and DYS385 a/b; ReliaGene Y-PLEX ™ 5 was used to type DYS389I/II, DYS392, DYS438, and DYS439; the Promega PowerPlex®Y prototype kit was used to type DYS19, DYS385 a/b, DYS389I/II, DYS390, DYS391, DYS392, DYS393, DYS437, DYS438, and DYS439; and the OligoTrail 20 plex prototype kit was used to type DYS19, DYS 385a/b, DYS389I/II, DYS 390, DYS391, DYS392, DYS393, YCAIIa/b, and DYS388. Three NIST multiplex assays [NIST 20 plex (27), NIST 11 plex (32,33) and NIST 10 plex (29)] were also used to examine the various Y STR markers.

Six components, labeled A through F, are included in each SRM unit and each component contains 50 μL of genomic DNA at a concentration of approximately 2 ng/μL. Components A through E are male genomic DNA 1 through 5, respectively. Component F is female genomic DNA, which serves as a negative control for Y-chromosome specific assays.

Table 14. Certified values for the European minimal haplotype and U.S. core Y-STR loci.

Component ID	Gender Name	DYS19	DYS389I	DYS389II	DYS390	DYS391	DYS392	DYS393	DYS385 a/b	DYS438	DYS439
A	Male 1	14	13	29	25	11	13	13	12-15	12	12
B	Male 2	14	13	28[a]	23	11	11	12	14-17	9	12
C	Male 3	16	14	32	21[b]	12	11	13	17-20	11	11
D	Male 4	15	12	28[a]	22	10	11	14	14-15	11	11
E	Male 5	17	14	31	24	10	12	14	13-15	10	11
F	Female	-	-	-	-	-	-	-	-	-	-

[a] In DYS389II, components B and D both contain a total number of repeats equal to 28, but can be distinguished since B has 5-10-3-10 and D has 5-11-3-9 of the TCTG-TCTA-TCTG-TCTA repeats, respectively (see Table 17).

[b] At the DYS390 locus, component C has a variant allele. Size-based analytical systems should type this sample as a 21 repeat allele although sequence information reveals a T-to-A polymorphism internal to the repeat.

Table 15. Certified values for other commonly used Y-STR markers

Component ID	Name	DYS 388	DYS 426	DYS 435	DYS 436	DYS 437	DYS 447	DYS 448	DYS 460	DYS 461	DYS 462	H4
A	Male 1	12	12	12	12	15	24	19[b]	11	12	11	12[c]
B	Male 2	15	11	11	12	14	25[a]	21[b]	10	13	11	12[c]
C	Male 3	12	11	11	12	14	25[a]	21[b]	9	13	12	12[c]
D	Male 4	12	11	11	12	16	23	21[b]	11	11	13	12[c]
E	Male 5	13	11	11	12	14	26	20[b]	11	12	12	11[c]
F	Female	-	-	-	-	-	-	-	-	-	-	-

[a] In DYS447, components B and C both have a total of 25 repeats. However, B contains 9-1-8-1-6, while C has 7-1-8-1-8 of the following [TAATA]-[TAAAA]-[TAATA]-[TAAAA]-[TAATA] repeats, repectively (see Table 17).

[b] Three different nomenclatures have been published for DYS448 (32). The one listed here follows Redd *et al.* (28).

[c] The H4 nomenclature follows Butler *et al.* (27).

Table 16. Informational values for additional Y-STR loci.

Component ID	Name	DYS 450	DYS 456	DYS 458	DYS464 a/b/c/d expanded	conservative	YCAII a/b
A	Male 1	10	15	16	14-15-15-17	14-15-17	19-23
B	Male 2	10	15	15	12-13-13-17	12-13-17	19-22
C	Male 3	8	15	17	13-16-16-18	13-16-18	19-19
D	Male 4	9	15	16	13-13-14-14	13-14	20-20
E	Male 5	10	15	16	11-14-14-15	11-14-15	19-21
F	Female	--	--	--	--	--	--

Table 17. Repeat sequences of Y-chromosome STRs

STR Loci	STR sequence
DYS19	$[TAGA]_X$ tagg $[TAGA]_Y$
DYS388	$[ATT]_X$
DYS391	$[TCTA]_X$
DYS392	$[TAT]_X$
DYS393	$[AGAT]_X$
DYS426	$[GTT]_X$
DYS435	$[TGGA]_X$
DYS436	$[GTT]_X$
DYS438	$[TTTTC]_X$
DYS439	$[GATA]_X$
DYS460	$[ATAG]_X$
DYS461	$[TAGA]_X$CAGA
DYS462	$[TATG]_X$
DYS385a	$[GAAA]_X$
DYS385b	$[GAAA]_X$
GATA-H4	$[TAGA]_X$
DYS3891	$[TCTG]_X[TCTA]_Y$
DYS38911	$[TCTG]_X[TCTA]_Y$ .... $[TCTG]_{X'}[TCTA]_{Y'}$
DYS390	$[TCTG]_X[TCTA]_Y[TCTG]_{X'}[TCTA]_{Y'}$*
DYS447	$[TAATA]_X$ [TAAAA]$[TAATA]_Y$[TAAAA]$[TAATA]_{X'}$
DYS448	$[AGAGAT]_X N_{42}[AGAGAT]_Y$
DYS437	$[TCTA]_X[TCTG]_Y[TCTA]_{X'}$

X + Y + X` + Y` = Total number of repeat units in each loci. For details, see SRM 2395 certificate of analysis and Table 4 at http://patapsco.nist.gov/srmcatalog/certificates/2395.pdf

* In male genomic DNA 3, one TCTA changed to ACTA. See Table 4 in the certificate of analysis.

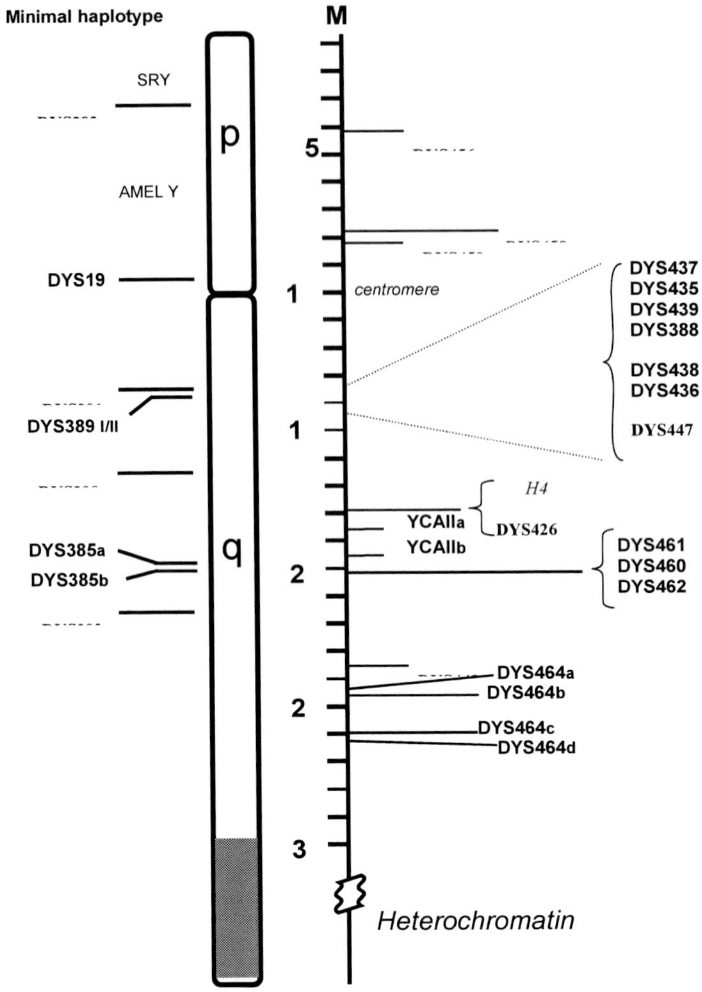

Figure 2. Y-STR Marker Positions for Loci in SRM 2395

Interlaboratory Evaluation of SRM 2395

The STR values for this SRM represent the pooled results from analyses performed at NIST, ReliaGene Technologies, Inc. (New Orleans, LA), OligoTrail LLC (Evanston, IL), and the Forensic Laboratory for DNA Research at Leiden University Medical Center (Leiden, The Netherlands). Sequencing of all (except DYS462) STR markers was performed at NIST. DYS462 was sequenced at Leiden University, who also confirmed the sequences of DYS19, DYS388, DYS389 I/II, DYS390, DYS391, DYS392, DYS393, DYS460, and DYS461.

In summary, the attributes of the Y-chromosome SRM 2395 can be compared to those of the mitochondrial DNA SRMs 2392 and 2392-I in that the Y-chromosome is inherited through the paternal lineage and the mtDNA is inherited through the maternal lineage. Thus, they both can be used to detect evolutionary backgrounds. However, they differ in that the Y-chromosome STR markers are only found in the male population; whereas, all individuals (males and females) have mtDNA that they have inherited from their mothers. If multiple loci are examined, the Y-chromosome STRs can be used to discriminate between individual males. The mtDNA will be similar among maternally-related family members whose DNA can be used to verify identification of missing children, unknown soldier remains, and criminal cases in which blood is found at the scene, but the suspect is not available. The mtDNA SRM can also be used to trace the inheritance of mitochondrial diseases and help in the diagnosis of diseases that have be correlated with mtDNA mutations. The Y-chromosome SRM 2395 will be especially helpful in distinguishing between the male and female DNA in sexual assault cases and in paternity disputes. All of the DNA SRMs that have been developed at NIST will serve to provide quality control and quality assurance that laboratories are testing their unknown samples correctly and obtaining accurate results.

ACKNOWLEDGMENTS

All of the analytical determinations and technical measurements leading to the certification of SRMs 2390, 2391, 2392, 2392-I, and 2395 were performed in the NIST Biotechnology Division of the Chemical Science and Technology Laboratory at the National Institute of Standards and Technology. The technical and support aspects involved in the preparation, certification and issuance of these SRMs were coordinated through the Standard Reference Materials Program by J. C. Colbert, C. S. Davis, and B. S. MacDonald.

For SRM 2390, the technical aspects of this recertification were conducted by D. J. Reeder, formerly of the NIST Biotechnology Division. Material stability studies and coordination of the interlaboratory study were conducted by K. L. Richie of the NIST Biotechnology Division. Data analysis was performed by D.L. Duewer of the NIST Analytical Chemistry Division. Statistical consultation was provided by H-K Liu of the NIST Statistical Engineering Division.

For SRM 2391b, the analytical determinations and technical measurements leading to the certification of this SRM were performed by M. C. Kline and J. W. Redman of the NIST Biotechnology Division. The overall direction of the technical measurement leading to the certification was under the chairmanship of J. M. Butler of the NIST Biotechnology Division. The DNA, amplified and genomic, was obtained from Roche Molecular Systems, Inc., Alameda, CA. The cell lines - GM09947A and GM09948 - were obtained from Marligen Biosciences, Inc., Ijamsville, MD.

For SRM 2392, the analytical determinations and technical measurements leading to the certification of this SRM were performed by B. C. Levin, H. Cheng, L. Tully, M. P. Jones and D. J. Reeder. The leader of the NIST DNA Technologies Group in the Biotechnology Division, D. J. Reeder, provided overall direction leading to the certification. DNA from GM09947A was obtained from Life Technologies, Inc. Funding was received from the National Institute of Justice through the NIST Office of Law Enforcement Standards and the Advanced Technology Program at NIST. P. Reddy provided help and advice on the cloning of the HV1 region.

For SRM 2392-I, the analytical determination, technical measurements and analysis of data for the certification of this SRM were performed by D. K. Hancock, K. L. Richie, K. A.

Holland (on sabbatical from Gettysburg College, Gettysburg, PA), and B. C. Levin of the NIST DNA Technologies Group, Biotechnology Division, Chemical Science and Technology Laboratory. The overall direction and coordination of the technical measurements leading to the certification was performed by B. C. Levin of the NIST DNA Technologies Group, Biotechnology Division, Chemical Science and Technology Laboratory. This research was supported by Interagency Agreement number 1999-IJ-R-094 from the National Institute of Justice.

For SRM 2395, the overall direction and coordination of the technical activities leading to certification were under the chairmanship of Margaret C. Kline and John M. Butler of the NIST Biotechnology Division. Analytical determination and technical measurements leading to the certification of this SRM were performed by J.M. Butler, R. Schoske, P.M. Vallone, M.C. Kline, and J.W. Redman of the NIST Biotechnology Division. This research was supported by Interagency Agreement number 1999-IJ-R-094 from the National Institute of Justice.

For SRMs 2394 and 2399, which were developed and released following the submission of this chapter, more information can be obtained from The National Institute of Standards and Technology, Standard Reference Materials Program, 100 Bureau Drive, Mail Stop 2322, Gaithersburg, MD 20899-2322, Telephone: (301) 975-6776, FAX: (301) 948-3730

E-mail: srminfo@nist.gov

Web sites: www.nist.gov/srm, http://ts.nist.gov/ts/htdocs/230/232/232.htm.

REFERENCES

[1] Reeder, D.J., NIST Standards Development for RFLP DNA Profiling. *Proceedings of the Second International Symposium on Human Identification, Promega Corporation*, p. 245-261 (1991).

[2] National Institute of Standards and Technology, "Certificate of Analysis, *Standard Reference Material* 2390, DNA Profiling Standard," Standard Reference Materials Program, NIST, Gaithersburg, MD (1992, 2001). (http://patapsco.nist.gov/srmcatalog/certificates/2390.pdf)

[3] Reeder, D.J., Kline, M.C., and Richie, K.L., An Overview of Reference Materials Prepared for Standardization of DNA Typing Procedures. Fresenius' *J. Anal. Chem.* 352:246-249 (1995).

[4] Levin, B.C., Cheng, H., Kline, M.C., Redman, J.W., and Richie, K.L., A Review of the DNA Standard Reference Materials Developed by the National Institute of Standards and Technology. Fresenius' *J. Anal. Chem.* 370:213-219 (2001).

[5] Duewer, D.L., Currie, L.A., Reeder, D.J., Leigh, S.D., Liu, H-K., and Mudd, J.L., Interlaboratory Comparison of Autoradiographic DNA Profiling Measurements. 4. Protocol Effects, *Anal. Chem.*, 69:1882-1892, (1997).

[6] Eisenberg, A.J., Gibson, P., Nandi, S., and Wang, L., The Development and Implementation of a Hae III Based RFLP System for Parentage Testing in Texas. *Proceedings from the Second International Symposium on Human Identification*, Promega Corporation, p. 163-180 (1991).

[7] Budowle, B., The RFLP Technique. *Crime Laboratory Digest* 15:97-98 (1988).

[8] Budowle, B. and Baechtel, F.S. Modifications to Improve the Effectiveness of Restriction Fragment Length Polymorphism Typing, Applied Theories. *Electrophoresis* **1**: 181-187 (1990).

[9] Hahn, G.J. and Meeker, W.Q. *Statistical Intervals: a Guide for Practitioners*. John Wiley, New York, NY (1991).

[10] Nakamura, Y., Carlson, M., Krapcho, V., and White, R., Isolation and Mapping of a polymorphic DNA Sequence (pMCT118) on Chromosome 1p (D1S80). *Nucleic Acids Res*. 16: 9364 (1988).

[11] Kasai, K., Nakamura, Y., and White, R., Amplification of a Variable Number of Tandem repeats (VNTR) Locus (pMCT118) by the Polymerase Chain Reaction (PCR) and its Application to Forensic Science. *J. For. Sci*. 35: 1196-1200 (1990).

[12] Anderson, S., Bankier, A. T., Barrell, B. G., deBrujin, M. H. L., Coulson, A. R., Drouin, J., Eperon, I. C., Nierlich, D. P., Roe, B. A., Sanger, F., Schreier, P. H., Smith, A. J. H., Staden, R., and Young, I. G., Sequence and Organization of the Human Mitochondrial Genome. *Nature* 290: 457-465 (1981).

[13] Andrews, R.M., Kubacka, I., Chinnery, P.F., Lightowlers, R.N., Turnbull, D.M., Howell, N. (1999). Reanalysis and revision of the Cambridge Reference Sequence for human mitochondrial DNA. *Nature Genetics* 23:147 (1999).

[14] Bogenhagen, D., and Clayton, D. A., The Number of Mitochondrial Deoxyribonucleic Acid Genomes in Mouse L and Human HeLa Cells. *J. Biol. Chem*. 249: 7991-7995 (1974).

[15] King, M. P., and Attardi, G., Human Cells Lacking MtDNA: Repopulation with Exogenous Mitochondria by Complementation. *Science* 246: 500-503 (1989).

[16] Holland, M. M., Fisher, D. L., Mitchell, L. G., Rodriquez, W. C., Canik, J. J., Merril, C. R., and Weedn, V. W., Mitochondrial DNA Sequence Analysis of Human Skeletal Remains: Identification of Remains from the Vietnam War. *J. Forensic Sciences* 38: 542-553. (1993).

[17] Holland, M. M., Fisher, D. L., Roby, R. K., Ruderman, J., Bryson, C., and Weedn, V. W., Mitochondrial DNA Sequence Analysis of Human Remains. *Crime Laboratory Digest* 22: 109-115 (1995).

[18] Wallace, D. C., Mitochondrial Genetics: a Paradigm for Aging and Degenerative Diseases. *Science* 256: 628-632 (1992).

[19] Grossman, L. I., Mitochondrial Mutations and Human Disease. *Env. & Mol. Mutag*. 25: (supplement 26) 30-37 (1995).

[20] Ballinger, S. W., Bouder, T. G., Davis, G. S., Judice, S. A., Nicklas, J. A., and Albertini, R. J., Mitochondrial Genome Damage Associated with Cigarette Smoking. *Cancer Research* 56: 5692-5697 (1996).

[21] Levin, B.C., Cheng, H., and Reeder, D.J., A Human Mitochondrial DNA Standard Reference Material for Quality Control in Forensic Identification, Medical Diagnosis, and Mutation Detection. *Genomics* 55: 135-146 (1999).

[22] Levin, B.C., Holland, K.A., Hancock, D.K., Coble, M., Parsons, T.J., Kienker, L.J., Williams, D.W., Jones, MP, and Richie, K.L., Comparison of the Complete mtDNA Genome Sequences of Human Cell Lines - HL-60 and GM10742A - from Individuals

with Pro-myelocytic Leukemia and Leber Hereditary Optic Neuropathy, Respectively, and the Inclusion of HL-60 in the NIST Human Mitochondrial DNA Standard Reference Material - SRM 2392-I. *Mitochondrion* **2**:387-400 (2003).

[23] Levin, B.C., Hancock, D.K., Holland, K.A., Cheng, H., and Richie, K.L., Human Mitochondrial DNA - *Amplification and Sequencing Standard Reference Materials* – SRM 2392 and SRM 2392-I. NIST Special Publication 260-155, National Institute of Standards and Technology, Gaithersburg, MD (2003).

[24] Jobling, M.A., Pandya, A., and Tyler-Smith, C. The Y-Chromosome in Forensic Analysis and Paternity Testing. *Int. J Legal Med.* 110:118-124 (1997).

[25] Kayser, M. *et al.* Evaluation of Y-Chromosomal STRs: a Multicenter Study. *Int. J. Legal Med.*110:125-133 (1997).

[26] Roewer, L. *et al.* Online Reference Database of European Y-Chromosomal Short Tandem Repeat (STR) Haplotypes. *Forensic Sci. Int.* 118:106-113 (2001).

[27] Butler, J.M., Schoske, R., Vallone, P.M., Kline, M.C., Redd, A.J., and Hammer, M.F. A Novel Multiplex for Simultaneous Amplification of 20 Y-Chromosome STR Markers. *Forensic Sci. Int.*129**:**10-24 (2002).

[28] Redd, A.J., Agellon, A.B., Kearney, V.A., Contreras, V.A., Karafet, T., Park, H., de Knijff, P., Butler, J. M., and Hammer, M. F. Forensic Value of 14 Novel STRs on the Human Y-Chromosome. *Forensic Sci. Int.*130**:** 97-111 (2002).

[29] Schoske, R., Vallone, P.M., Ruitberg, C.M., and Butler, J.M. Multiplex PCR Design Strategy Used for the Simultaneous Amplification of 10 Y-Chromosome Short Tandem Repeat (STR) Loci. *Anal. Bioanal. Chem*.375**:**333-343 (2003).

[30] Sinha, S. et al. Development and Validation of a Multiplexed Y-Chromosome STR Genotyping System, Y-PLEX. 6, for Forensic Casework. *J. Forensic Sci.* 48**:**93-103 (2003).

[31] Butler, J.M. Recent Developments in Y-STR and Y-SNP Analysis. *Forensic Sci. Rev.* (in press, 2003).

[32] Schoske, R. The Design, *Optimization and Testing of Y-Chromosome Short Tandem Repeat Megaplexes*; Ph.D. Dissertation, American University (2003).

[33] Schoske, R., Vallone, P.M., Kline, M.C., Redman, J.W., and Butler, J.M. High-Throughput Y-STR Typing of U.S. *Populations With 27 Regions of the Y-Chromosome Using Two Multiplex PCR Assays.* submitted (2003).

[34] Vallone, P.M., and Butler, J.M. Y-SNP Typing Using Allele-Specific Hybridization and Primer Extension. submitted (2003).

[35] Y Chromosome Consortium. A Nomenclature System for the Tree of Human Y-Chromosomal Binary Haplogroups. *Genome Res*.**7:**339-348 (2002).

In: DNA Research Trends
Editor: Marius P. Young, pp. 101-124
ISBN: 978-1-60021-646-6

Chapter IV

TISSUE ENGINEERING USING ADIPOSE-DERIVED STEM CELLS HARVESTED FROM GFP TRANSGENIC ANIMALS

***Rei Ogawa*[1,2,*], *Juri Fujimura*[3], *Hiroshi Mizuno*[1], *Hiko Hyakusoku*[1] and *Takashi Shimada*[2]**

[1]Department of Plastic and Reconstructive Surgery,
[2]Department of Biochemistry and Molecular Biology,
[3]Department of Pediatrics,
Nippon Medical School, Tokyo, Japan.

ABSTRACT

Adult stem cells are considerably easier to manipulate than either embryonic stem or embryonic germ cells. At present, the authors believe that of all the mesenchymal stem cells, adipose-derived stem cells (ASCs) are the most promising due to their pluripotency, proliferative efficiency, and low donor morbidity. In this chapter we describe experiments using ASCs both *in vitro* and *in vivo*, and in particular, we describe the importance of GFP transgenic mice and rats in these studies. We sought to determine if ASCs harvested from GFP transgenic animals could differentiate into osteogenic, chondrogenic, adipogenic, and neural lineages *in vitro*. In an *in vivo* study, we investigated bone and bone marrow regeneration using ASCs harvested from GFP transgenic mice. The results suggest that ASCs harvested from both GFP transgenic mice and rats have the potential to differentiate into osteogenic, chondrogenic, adipogenic, and neural lineages *in vitro*. Additionally, the results indicate that bone and bone marrow can be regenerated subcutaneously, using ASCs harvested from GFP transgenic mice.

* Correspondence concerning this article should be addressed to Rei Ogawa, MD, PhD, Department of Plastic and Reconstructive Surgery Nippon Medical School, 1-1-5 Sendagi Bunkyo-ku, Tokyo 113-8603, Japan. Phone: +81-3-5814-6208; Fax: +81-3-5685-3076; E-mail: r.ogawa@nms.ac.jp.

Introduction

Multipotent mesenchymal stem cells (MSCs) were first described in 1976 [1]. These adult stem cells are much easier to manipulate than either embryonic stem (ES) or embryonic germ (EG) cells. However, they do not demonstrate the "true" totipotency of ES/EG cells. The major advantage to using these cells is that there are few ethical or immunological problems associated with their use in auto transplantation. Therefore, expectations are high for the future use of auto transplantation of MSCs, in regenerating or repairing damaged or diseased tissues.

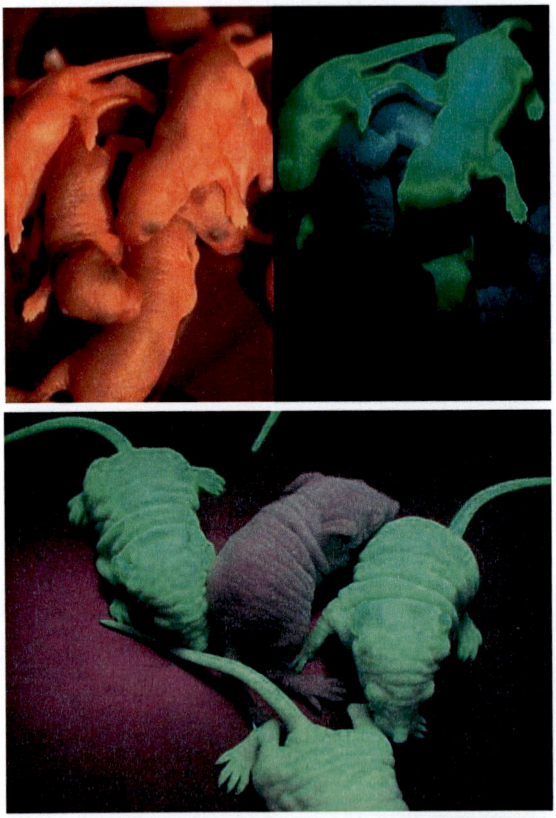
A

B

Figure 1. GFP transgenic mice. A: Newborn mice, B: Adult mice. The GFP transgenic mice were uniformly green under excitation light, with the exception of hair and red blood cells. These figures were provided by Dr. Okabe at the Genome Information Research Center, Osaka University, Osaka, Japan. Reproduced with permission of FEBS Letters. (Okabe, M., et al. (1997) 'Green mice' as a source of ubiquitous green cells. FEBS Lett, 407, 313-9.)

For a long time, bone marrow was the major source of MSCs; however, recent studies have identified new sources. For example, Zuk et al. [2] have reported MSCs in adipose tissue, Young et al. [3] found that they can be harvested from skeletal muscles and dermis, and Erices et al. [4] have demonstrated that they are present in umbilical cord blood. Further investigation of these MSCs is necessary, since at present, we do not know if all of these cells are the same, and there is the possibility that each may have characteristics that favor specific types of differentiation, depending on their source. The authors are of the opinion that adipose-derived stem cells (ASCs) are the easiest of the MSCs to manipulate due to their pluripotency, proliferative efficiency, and low donor morbidity.

Both green fluorescent protein (GFP) transgenic mice (Figure 1) produced in 1997 [5] and GFP transgenic rats (Figure 2) produced in 2001 [6] are considered to be useful for cell transplantation experiments. The fluorescent nature of the cells from these transgenic animals provides an indication of the cell source in *in vivo* studies, because cell populations can be easily identified through fluorescence microscopy. Thus, in this chapter, we also describe the animal experiments using ASCs both *in vitro* and *in vivo*, and discuss the importance of GFP transgenic mice and rats for studying ASCs. We sought to determine if ASCs harvested from GFP transgenic animals could also differentiate into osteogenic, chondrogenic, adipogenic, and neural lineages *in vitro*. Then, for the *in vivo* study, we investigated bone and bone marrow regeneration using ASCs harvested from GFP transgenic mice.

A

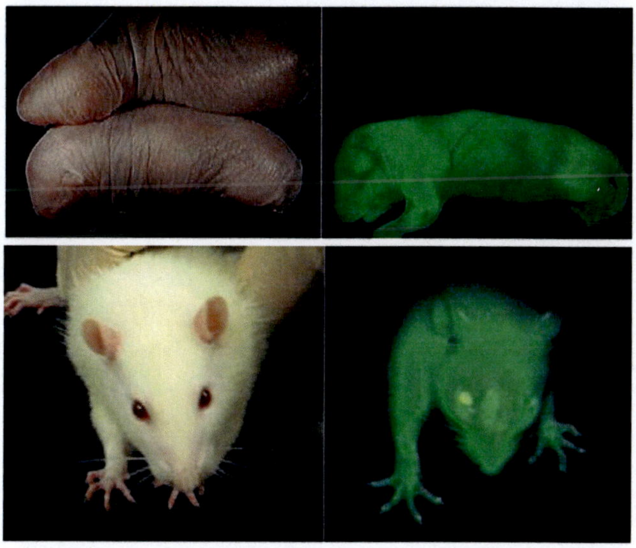

B

Figure 2. GFP transgenic rats. A: Newborn rats, B: Adult rats. The GFP gene introduced into rats was expressed almost ubiquitously in the organs and cells. The whole body of each newborn transgenic rat fluoresced green under excitation light. These figures were provided by Dr. Hakamata at the Center for Molecular Medicine, Jichi Medical School, Tochigi, Japan. Reproduced with permission of Biochemical Biophysical Research Communications. (Hakamata, Y., et al. (2001) Green fluorescent protein-transgenic rat: a tool for organ transplantation research. Biochem Biophys Res Commun, 286, 779-85.)

METHODS

In vitro Study

1. Cell harvest

Primary cultures were prepared with some modifications to previously described techniques [7,8]. All animals used for this study were cared for under the National Research Council's guidelines for the care and use of laboratory animals. Five-week-old GFP transgenic mice (C57BL/6 TgN [act-EGFP] OsbC14-Y01-FM131) [5] and GFP transgenic rats (Crj: Wister-TgN [CAG/GFP] 184ys) [6] were anesthetized with Nembutal (1 mg kg^{-1}). Inguinal fat pads were removed (1.0 g) and extensively washed with phosphate-buffered saline (PBS, Gibco BRL, Grand Island, N.Y.). They were then finely minced, and incubated in tissue culture plates (10 cm^2, Becton Dickinson, Franklin Lakes, N.J.) for 1 h in a control medium, Dulbecco modified Eagle medium (DMEM, Gibco BRL, Grand Island, N.Y.) with 10% fetal bovine serum (FBS, Gibco BRL, Grand Island, N.Y.), containing 30% antibiotic-antimycotic (Gibco BRL, Grand Island, N.Y.). The tissue was rinsed three times for 5 min in PBS, digested with 0.1 % (w/v) collagenase (Wako, Osaka) while vigorously shaking for 30 min at 37 °C, and an equal volume of control medium was added to neutralize the collagenase. The cell suspension was then centrifuged at 1500 rpm (240 *g*) for 5 min in a 50 mL centrifuge tube, and the cell pellet was resuspended in control medium. After cell counting using trypan blue, viable cells were plated at a concentration of 10^5 cells per plate (10 cm^2) and maintained in the control medium at 37 °C and 5 % CO_2.

2. Osteogenic Differentiation

The primary culture was expanded in the control medium for three passages, and cells were then trypsinized and replated at a concentration of 10^5 cells per plate (10 cm^2). These cells were incubated in the control medium for a day, to adhere them to the plates. The medium was then replaced by osteogenic medium (DMEM, 10 % FBS, 0.1 uM dexamethasone [Sigma, St. Louis, MO], and 50 ug mL^{-1} ascorbate-2-phosphate [Sigma, St. Louis, MO], 10mM beta-glycero-2-phosphate [Sigma, St. Louis, MO]) [8]. Culture media were changed every 3 days. Osteogenic differentiation was assessed by electron micrographs, von Kossa and alkaline phosphatase staining, starting a week after the initial osteogenic induction. Expression of genes related to osteogenic cells, such as alkaline phosphatase, osteopontin and osteocalcin, were confirmed by cDNA microarray and RT-PCR, as described previously [9].

3. Chondrogenic Differentiation

The primary culture was expanded in the control medium for three passages, and cells were then trypsinized and replated at a concentration of 10^7 cells per plate (10 cm^2). Chondrogenic differentiation was induced using a modified high-density "micromass" culture, as described previously [10]. 10 uL drops of cellular suspension were placed at 1 cm intervals on each plate. The plates were placed in an incubator at 37 °C with 5 % CO_2 for 5 h, then the medium was replaced by chondrogenic medium (DMEM, 1 % FBS, 10 ng mL^{-1} transforming growth factor-beta 1 [Sigma, St. Louis, MO], 6.25 ug mL^{-1} insulin [Sigma, St. Louis, MO], and 6.25 ug mL^{-1} transferrin [Sigma, St. Louis, MO]) [8,9]. Culture media were changed every 3 days. Chondrogenic differentiation was assessed by electron micrographs and Alcian blue staining at pH 1.0, starting a week after the initial chondrogenic induction. Expression of genes related to chondrogenic cells, such as aggrecan and type II/X collagen, were also confirmed by cDNA microarray and RT-PCR, as described previously [9].

4. Adipogenic Differentiation

The primary culture was expanded in the control medium for three passages, after which cells were trypsinized and replated at a concentration of 5 X 10^6 cells per plate (10 cm^2). These cells were incubated in the control medium for a day, to adhere them to the plates, after which the medium was replaced by adipogenic medium (DMEM, 10% FBS, 1 uM dexamethasone [Sigma, St. Louis, MO], 10 nM insulin [Wako, Osaka], 200 uM indomethacin [Sigma, St. Louis, MO] and 0.5 mM isobuthyl-methylxanthine [Sigma, St. Louis, MO]) [7]. Culture media were changed every 3 days. Adipogenic differentiation was assessed by electron micrographs and Oil Red O staining, starting a week after initial adipogenic induction. Expression of genes related to adipogenic cells, such as peroxisome proliferative activated receptor gamma2 (PPAR-gamma2) were confirmed by cDNA microarray and RT-PCR, as described previously [11].

5. Neural Differentiation

The primary culture was expanded in the control medium for three passages, and cells were then trypsinized and replated at a concentration of 10^6 cells per plate (10 cm^2). At 40-60 % confluence, the medium was replaced by a neural induction medium, prepared according to a previous report, but without FBS (DMEM, 1 % antibiotic-antimycotic, 5 ug mL^{-1} insulin, 200 uM indomethacin, and 0.5 mM isobutylmethylxanthine) [12]. At several time points from 6 to 48 h following neural induction, the cells were fixed and assessed by electron micrographs and immunocytochemistry, using the following mouse monoclonal antibodies: anti-nestin, anti-microtubule-associated protein 2 (MAP 2a + 2b), anti-70-kDa neurofilament protein (NF-70), and anti-glial fibrillary acidic protein (GFAP). Neural cell-related gene expression was confirmed by cDNA microarray and RT-PCR, as described previously [13].

In vivo Study

1. Cell harvest

Ten GFP transgenic mice (C57BL/6 TgN [act-EGFP] OsbC14-Y01-FM131), aged 5 weeks, were used for this study. The method of primary culture is described above.

2. Osteogenic Induction and Transplantation of the Cells

The primary culture was expanded in the control medium for three passages, and 10^6 cells were trypsinized and replated onto a CELLYARD (Pentax, Tokyo) hydroxyapatite scaffold, containing small pores (2–200 um diameter). Cells were incubated in an osteogenic medium (DMEM, 10 % FBS, 0.1 uM dexamethasone, 50 uM ascorbic acid-2-phosphate, and 10 mM beta-glycerophosphate [Sigma, St. Louis, MO]) for three days, to induce osteogenesis. Twenty five-week-old, GFP-negative, immunocompetent mice (C57BL/6), that had genetic homology to the GFP transgenic mice, were anesthetized with pentobarbital sodium at 1 mg kg^{-1}. ASC-containing scaffolds were transplanted subcutaneously onto the right back of each mouse. As controls, scaffolds not containing cells were transplanted subcutaneously onto the left back of each mouse.

3. Evaluation of bone and bone marrow Regeneration

From 4 weeks to 10 weeks after the operation, each 5 mice were anesthetized with 1 mg kg^{-1} sodium pentobarbital, every 2 weeks. X-ray images were taken, and the scaffolds removed for visual observation and histopathological evaluation. Scaffolds with surrounding tissues were fixed for 4 days and decalcified for 7 days with 4 % (w/v) paraformaldehyde (Sigma, St. Louis, MO) and 7 % (w/v) formic acid. Samples were embedded in paraffin and sectioned with a microtome to a thickness of 4 um. They were floated in water at 46 °C, placed on microscope slides, then incubated at 37 °C for 24 h. Histopathological and immunohistopathological evaluations of the specimens were performed using hematoxylin-eosin staining and anti-GFP immunostaining. As green florescence was not confirmed directly after making paraffin sections, anti-GFP immunostaining was performed, to confirm the presence of donor cells. For each staining procedure, the sections were dewaxed in xylene and dehydrated in an ethanol bath. Samples for histopathological staining were stained in hematoxylin for 20 min and in eosin for 5 min. Immunostaining was performed with an anti-GFP antibody (Cosmo Bio, Tokyo), a Vectastain ABC kit (Funakoshi, Tokyo), and counterstaining was carried out with hematoxylin.

RESULTS

In vitro Study

1. Cell Harvest

The populations of primary cultured cells, harvested from both GFP transgenic mice and rats, were heterogeneous and included a fibroblast-like cell morphology, as described previously [7-9] (Figure 3A). All cells concomitantly expressed GFP fluorescence under fluorescence microscopy (Figure 3B). Microscopically, subcultured cells appeared to be homogeneously fibroblastic after two passages. Under electron microscopy, almost all of the ASCs were larger than 20 um, and histological evaluation showed the cytoplasmic organelles to be relatively developed (Figure 3C). Moreover, there were few chromatin granules accumulated around the nuclei. Additionally, differences between the ASCs from mice and rats were found: those harvested from rats had both myosin and actin, whereas those from mice had only actin.

2. Osteogenic Differentiation

ASCs harvested from both GFP transgenic mice and rats, and placed in osteogenic media, exhibited changes in cell structure after 5 days in culture (Figure 4A). Up to 50 % of the total cell population changed from an elongated fibroblastic appearance, to a rounder, more cuboidal shape. All cells concomitantly expressed strong GFP fluorescence (Figure 4B). After one week of osteogenic induction, islands of extracellular matrices were secreted, vacuolation was observed under electron microscopy (Figure 4C), cells stained positively for endogenous alkaline phosphatase activity (Figure 5A), and mineralized modular structures were identified by von Kossa staining (Figure 5B). Additionally, the cells were examined by RT-PCR and cDNA microarray, and we confirmed the increased expression of several osteogenic-related genes, including an early marker of osteogenesis, alkaline phosphatase, as well as genes encoding the bone matrix proteins osteopontin and osteocalcin.

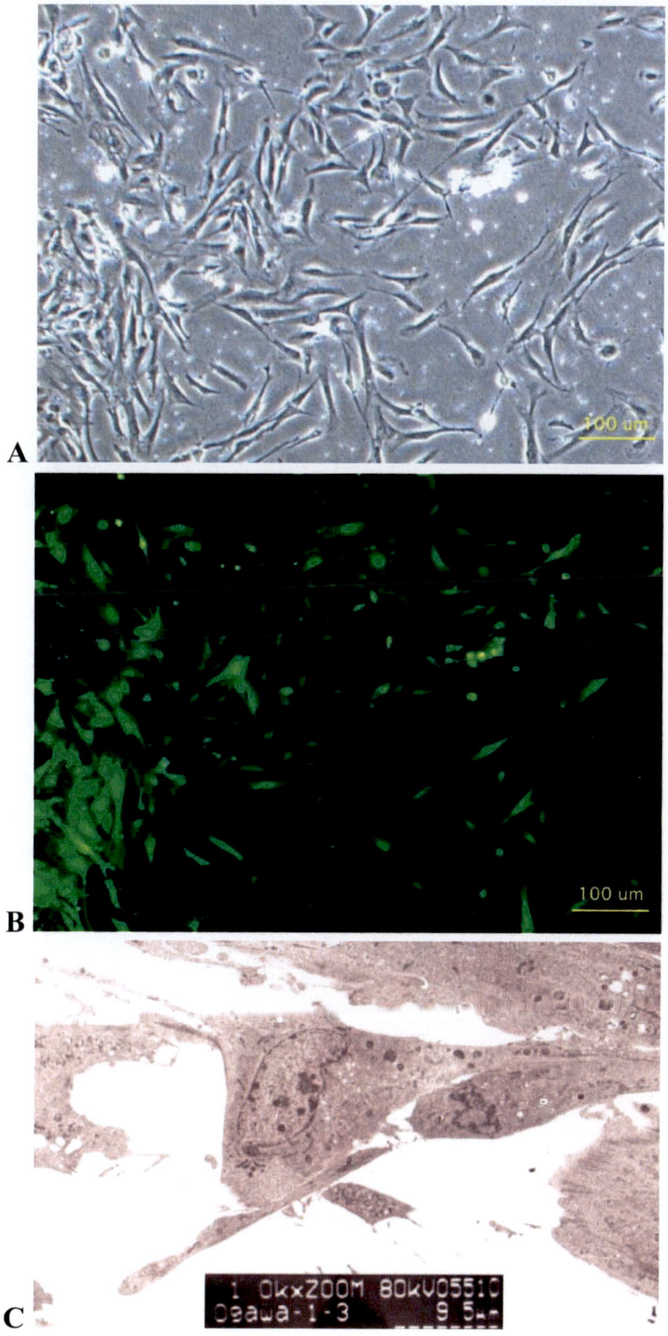

Figure 3. Adipose-derived stem cells (passage 2 cells isolated from GFP transgenic mice). A: Light microscope, B: Fluorescence microscope, C: Electron microscope. Adipose-derived stem cells isolated from both GFP transgenic mice and rats, expressing GFP fluorescence. The cells expressed GFP continuously after culture passage. Electron microscopic view shows few chromatin granules accumulated around the nuclei.

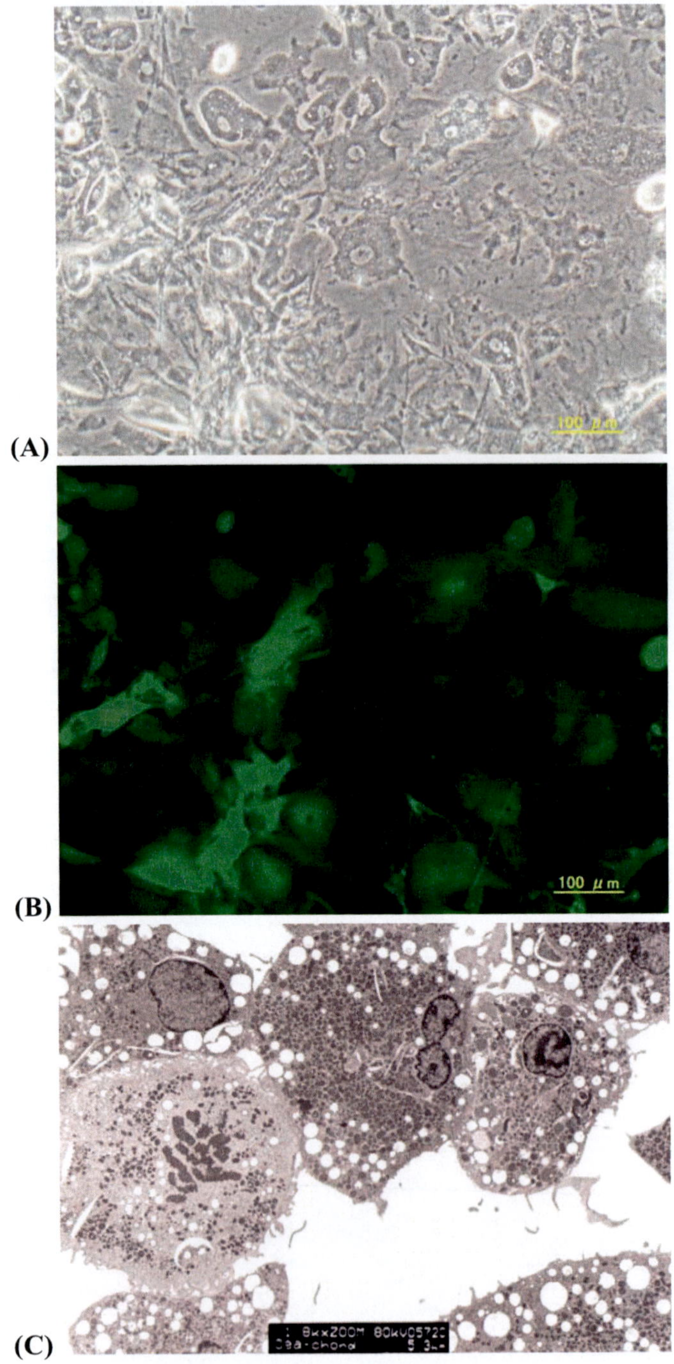

Figure 4. Adipose-derived stem cells incubated in osteogenic media (day-30 osteogenic induced cells harvested from GFP transgenic mice). A: Light microscope, B: Fluorescence microscope, C: Electron microscope. In osteogenic media, the cells changed from an elongated fibroblastic appearance to a rounder shape. Cells concomitantly expressed GFP fluorescence, and vacuolation was observed under electron microscopy.

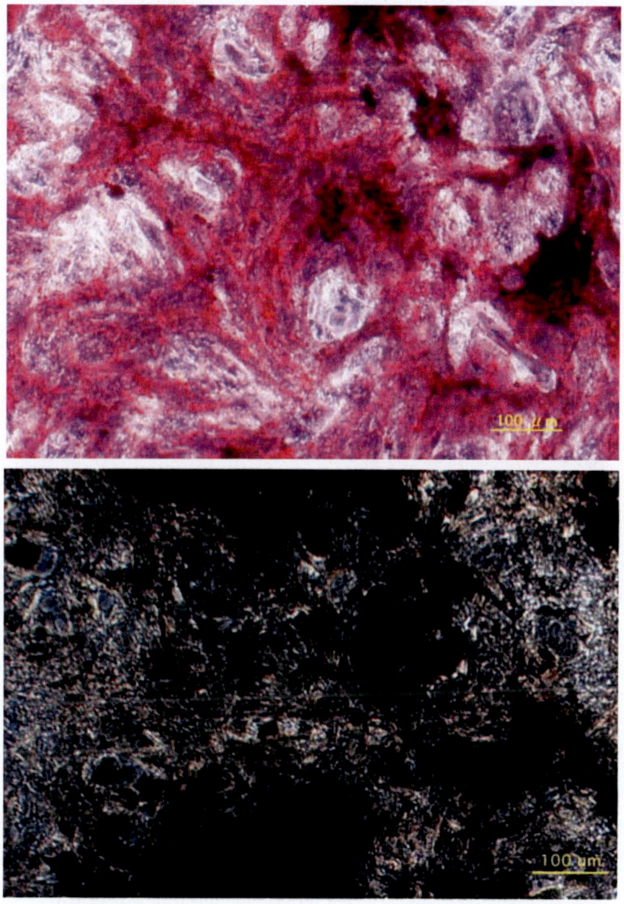

A

B

Figure 5. Specific staining of adipose-derived stem cells incubated in osteogenic media. A: Alkaline phosphatase staining (Day-7 osteogenic induced cells harvested from GFP transgenic mice), B: von Kossa staining (Day-30 osteogenic induced cells harvested from GFP transgenic mice). After a week of culture in osteogenic media, the cells stained positively for endogenous alkaline phosphatase activity, as well as for mineralized modular structures, confirmed by von Kossa staining.

.3. Chondrogenic Differentiation

High-density micromass cultures of ASCs harvested from both GFP transgenic mice and rats, and incubated in chondrogenic media, condensed into small spheroids that were visible to the naked eye as early as 3 days following chondrogenic induction (Figure 6A). Nodules concomitantly expressed strong GFP fluorescence (Figure 6B) and over the course of 10 days, the number of cartilaginous nodules increased. No nodules were observed in monolayer cultures prior to being placed in chondrogenic media. A layer of cells around the nodule was observed under electron microscopy, and the cells were aggregated by secreted collagen fibers (Figure 6C). Micromass cultured cells that had been supplemented with chondrogenic media, stained positively for Alcian blue at pH 1.0, which is specific for the highly sulfated proteoglycans of cartilage matrices (Figure 7). Additionally, the cells supplemented with chondrogenic media were examined by RT-PCR and cDNA microarray and we confirmed differential expression of aggrecan and type II/X collagen, as compared with control cells incubated in control media.

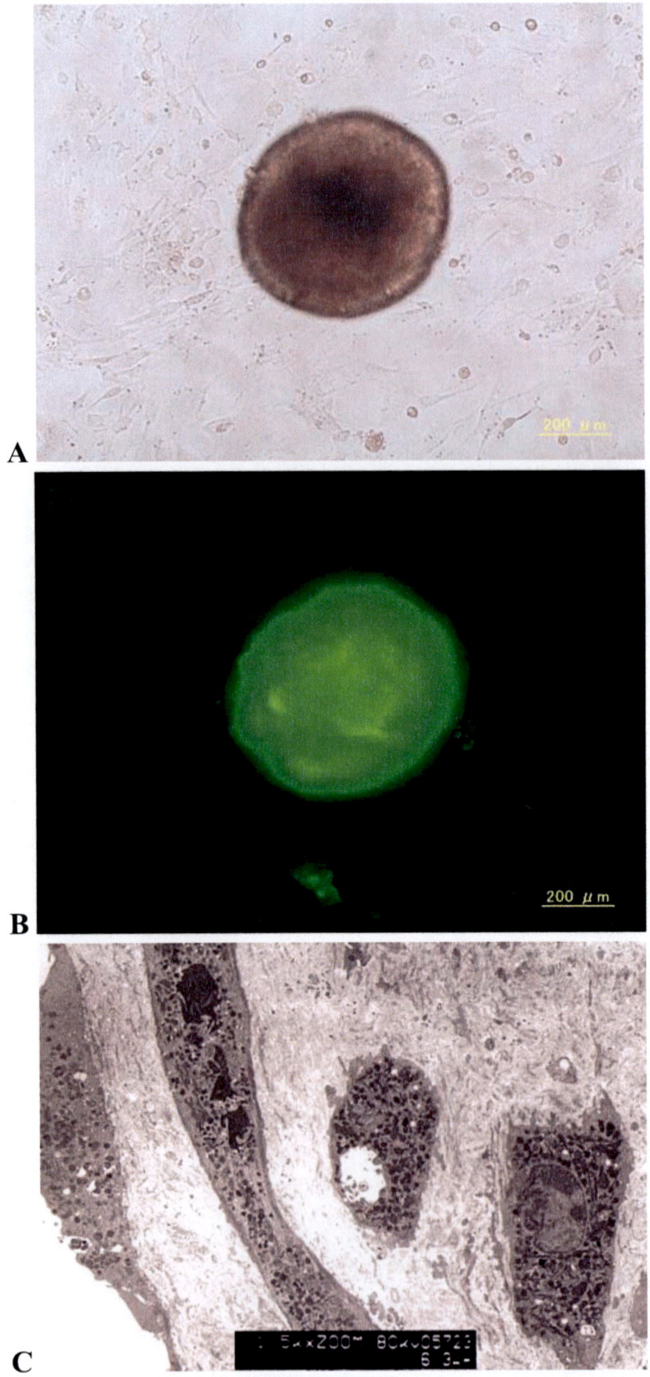

Figure 6. Adipose-derived stem cells incubated in chondrogenic media (day-14 chondrogenic induced cells harvested from GFP transgenic mice). A: Light microscope, B: Fluorescence microscope, C: Electron microscope. High-density micromass cultures of ASCs were harvested from both GFP transgenic mice and rats, and incubated in chondrogenic media. They condensed into small spheroids that were visible to the naked eye and that expressed strong GFP fluorescence. A layer of cells around the nodule was visible under electron microscopy, and the cells were seen to be aggregated via secreted collagen fibers.

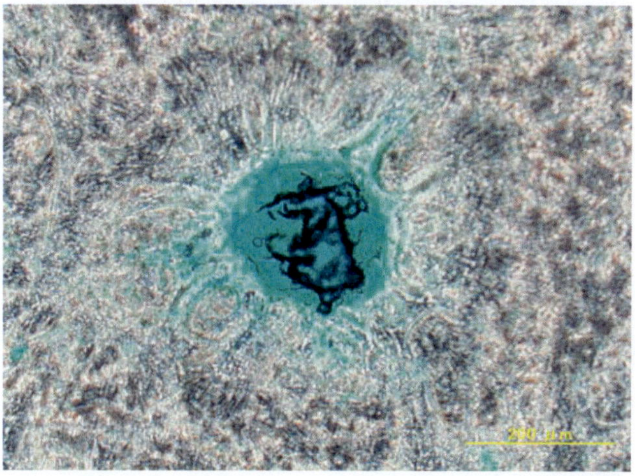

Figure 7. Alcian blue staining of adipose-derived stem cells incubated in chondrogenic media (day-14 chondrogenic induced cells harvested from GFP transgenic mice). Micromass cultured cells supplemented with chondrogenic media and stained positively with Alcian blue at pH 1.0. Staining is specific to the highly sulfated proteoglycans of cartilage matrices.

4. Adipogenic Differentiation

ASCs harvested from both GFP transgenic mice and rats, and incubated in adipogenic media, exhibited changes in cell structure after 7 days in culture. Up to 40 % of the total cell population changed from an elongated fibroblastic appearance, to a rounder shape (Figure 8A). All cells concomitantly expressed GFP fluorescence (Figure 8B). Two weeks after adipogenic induction, cells containing neutral lipid droplets were observed under electron microscopy (Figure 8C), and stained positively with the lipid dye, Oil Red O (Figure 9). Additionally, the cells supplemented with adipogenic media were examined by RT-PCR and cDNA microarray, and we confirmed increased expression of the adipogenic-related gene PPAR-gamma2, as compared with control cells incubated in control media.

5. Neural Differentiation

The morphology of ASCs harvested from both GFP transgenic mice and rats started to change within a few hours of neural induction (Figure 10A). All cells concomitantly expressed GFP fluorescence and developed characteristic round cell bodies, with several branching extensions (Figure 10B). Up to 60 % of the total cell population was bipolar or multipolar in shape and some of them appeared to make contact with neighboring cells. Following neural induction, many microtubules and thinner actin filaments were observed in the soma and processes of the multipolar cells (Figure 10C, D). The microtubules were long, smooth-walled hollow cylindrical structures, several micrometers long and around 20 nm in diameter, and occurred frequently in the dendrites and axons of neurally-induced cells. Most of the ASCs with bipolar processes expressed nestin, while those with multipolar processes tended to be stained with MAP2 or neurofilament-70, but not nestin. Additionally, there were some GFAP-positive cells. At present, in cells undergoing neural induction, we have confirmed increased expression of several neurally-related genes, including MAP2, as compared with control cells incubated in control media.

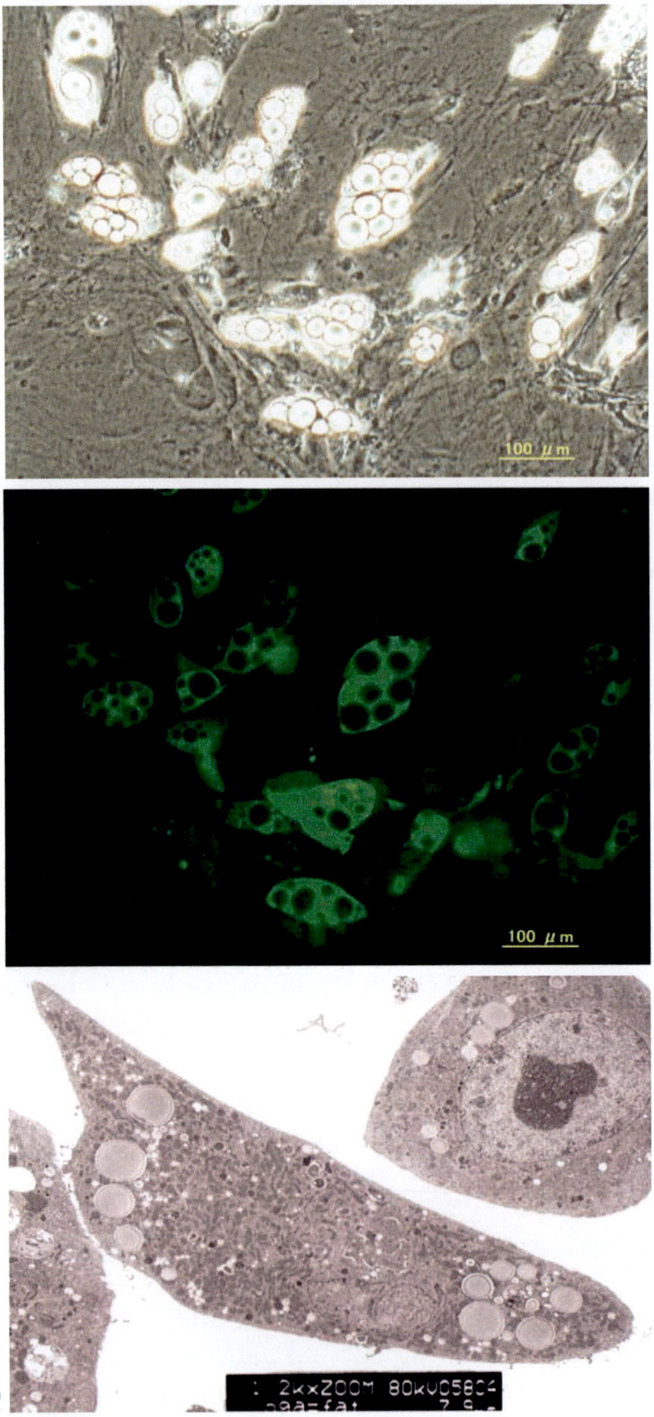

Figure 8. Adipose-derived stem cells incubated in adipogenic media (day-14 adipogenic induced cells harvested from GFP transgenic mice). A: Light microscope, B: Fluorescence microscope, C: Electron microscope. Cells harvested from both GFP transgenic mice and rats incubated in adipogenic media exhibited changes in cell structure after 7 days in culture, with expression of GFP fluorescence. In our electron microscopy study, neutral lipid droplets were detected in the cells.

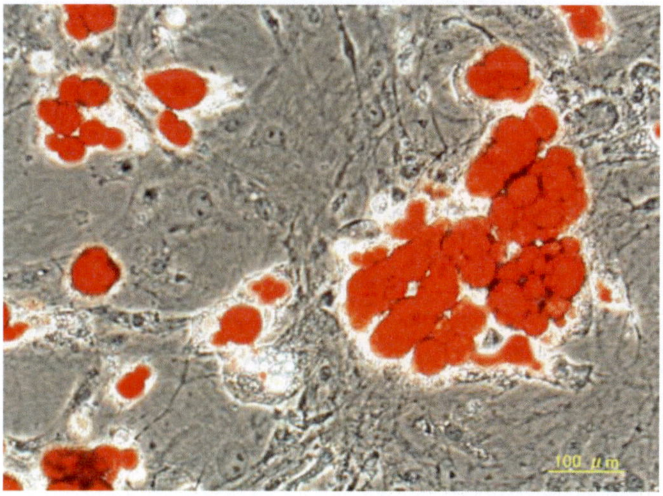

Figure 9. Oil Red O staining of adipose-derived stem cells incubated in adipogenic media (day-14 adipogenic induced cells harvested from GFP transgenic mice). Cell staining was positive with the lipid dye, Oil Red O.

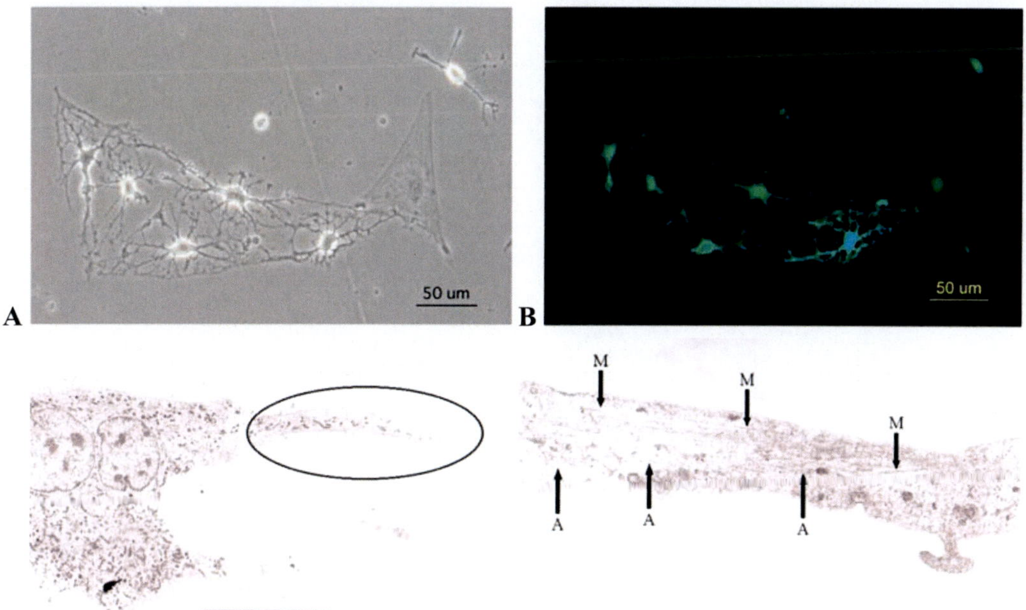

Figure 10. Neural induced adipose-derived stem cells (24 hours after neural induction harvested from GFP transgenic mice). A: Light microscope, B: Fluorescence microscope, C: Electron microscope (low power), D: Electron microscope (high power). After neural induction, the morphology of ASCs harvested from both GFP transgenic mice and rats began to change within a few hours. The cells developed characteristic round bodies, with several branching extensions, and concomitantly expressed GFP fluorescence. The multipolar cells exhibited many microtubules and thinner actin filaments in their soma and processes.

In vivo Study

1. Cell Harvest

As described in the in vitro study above, the populations of primary cultured cells, harvested from both GFP transgenic mice and rats, were heterogeneous, and included a fibroblast-like cell morphology.

2. Osteogenic Induction and Transplantation of Cells

After 3 passages in control medium, subcultured cells were homogeneously fibroblastic and exhibited bright green fluorescence under fluorescence microscopy. This cell population was plated onto scaffolds and induced in osteogenic medium for three days. At that point, we confirmed cell attachment to the scaffolds, as the scaffolds appeared bright green under fluorescence microscopy. These scaffolds were transplanted subcutaneously into immunocompetent GFP-negative mice (C57BL/6), which were genetically homologous to the GFP transgenic mice. Incision wounds were cured within a week, and no subsequent infection occurred during the course of their captivity.

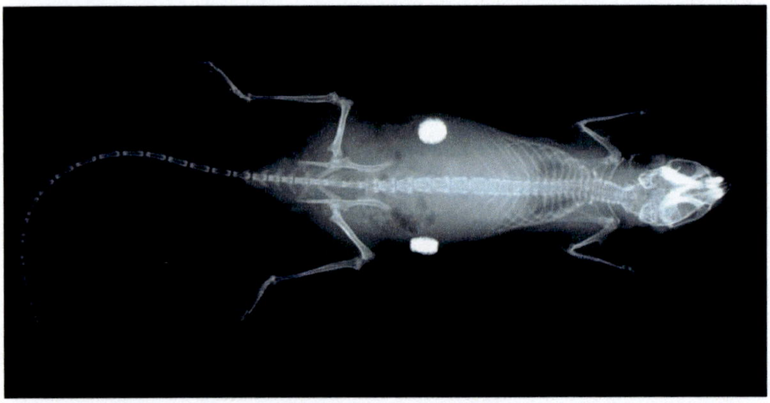

Figure 11. X-ray film of transplanted scaffolds containing adipose-derived stem cells harvested from GFP transgenic mice (4 weeks after transplantation). Scaffolds made of hydroxyapatites, containing osteogenic induced adipose-derived stem cells harvested from GFP transgenic mice, were transplanted subcutaneously into GFP-negative mice (C57BL/6) that were immunocompetent and genetically homologous to GFP transgenic mice. On X-ray film, the size of the scaffolds did not change during the 10 week period, regardless of the presence or absence of cells.

3. Evaluation of Bone Regeneration

Scaffolds were observed as shadows on X-ray film from 4 to 10 weeks following transplantation (Figure 11). The size of experimental and control scaffolds did not change during this time. At each time point at which scaffolds were extirpated from the recipient mice, large numbers of infiltrating microvessels were observed on scaffolds containing ASCs. However, on control scaffolds, few microvessels were observed (Figure 12A). Under fluorescence microscopy, cells on the scaffolds were bright green and were found to be particularly concentrated in the micropores (Figure 12B). However, GFP expression reduced over time and 3 of the 5 specimens removed after 10 weeks did not express any fluorescence.

The other 2 of the 5 specimens expressed a low level of fluorescence. No fluorescence was observed on any of the control scaffolds.

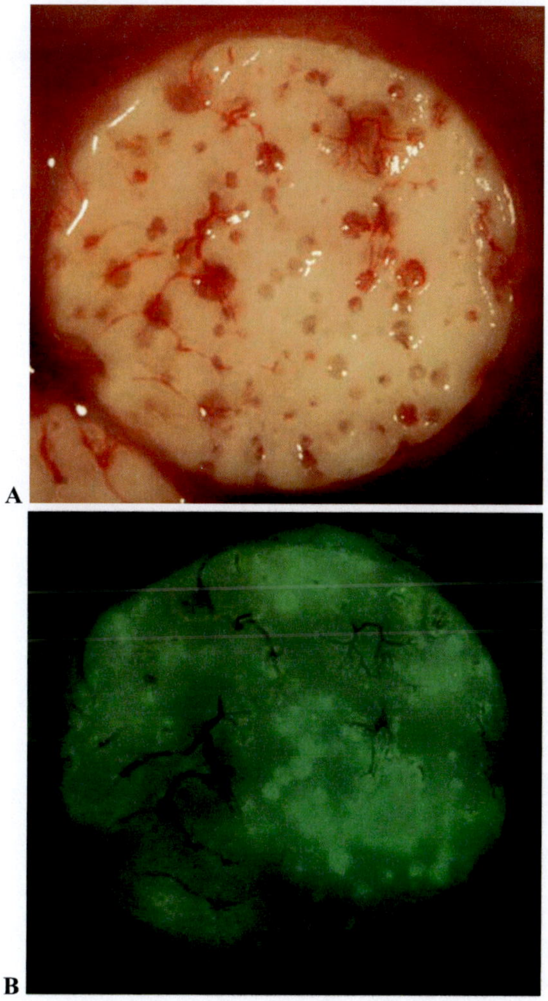

Figure 12. Scaffolds containing adipose-derived stem cells harvested from GFP transgenic mice (4 weeks after transplantation). A: Light microscope, B: Fluorescence microscope. Large numbers of infiltrating microvessels were observed. Under the fluorescence microscope, the cells on the scaffolds, especially in the micropores, were a strong fluorescent green. No fluorescence was observed in control scaffolds.

Under microscopic observation of hematoxylin-eosin stained specimens of scaffolds containing ASCs, we observed bone matrix and bone marrow, adipocytes, a variety of blood cells, and small vessels, in the scaffold micropores (Figure 13). At 8 and 10 weeks after transplantation, we observed more developed bone matrix and bone marrow structures than had been observed for specimens obtained at 4 and 6 weeks. In all control scaffolds, only fibrous structures and inflammatory cells were observed; neither bone matrix nor bone marrow were present.

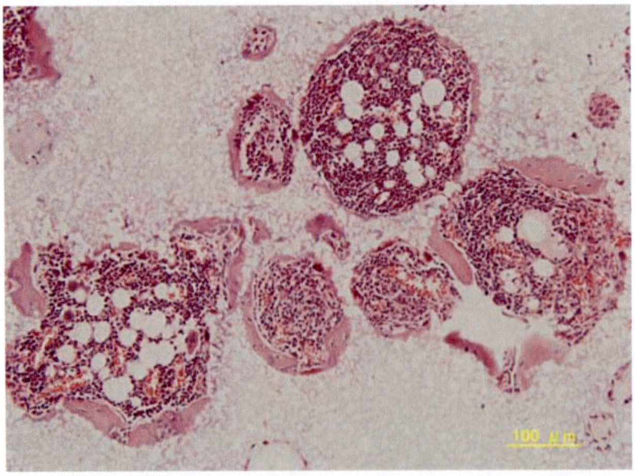

Figure 13. A hematoxylin-eosin stained specimen of a scaffold containing adipose-derived stem cells (4 weeks after transplantation). Bone matrix and bone marrow in micropores surrounded by hydroxyapatite particles, including adipocytes, various kinds of blood cells, and small vessels. Only fibrous structures and inflammatory cells were observed in control scaffolds, and neither bone matrix nor bone marrow were present.

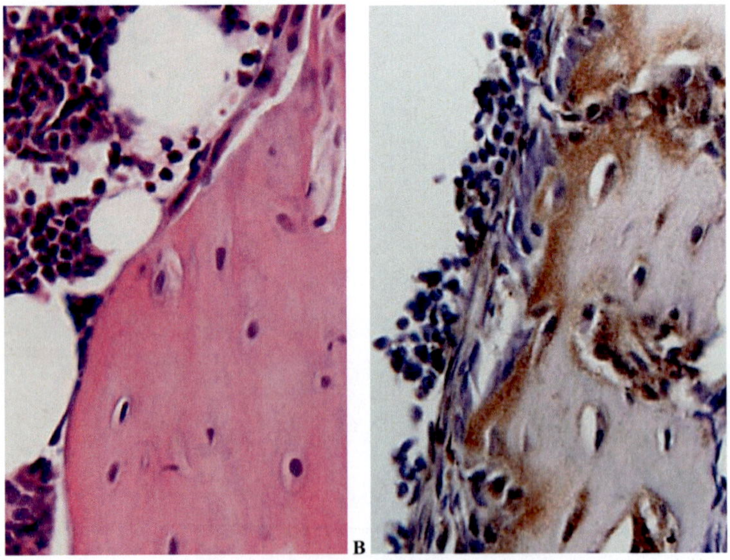

Figure 14. An anti-GFP immunostained specimen of a scaffold containing adipose-derived stem cells (4 weeks after transplantation). A: Hematoxylin-eosin staining, B: Anti-GFP immunostaining. Since green florescence was not confirmed directly after making paraffin sections, anti-GFP immunostaining was performed to confirm the presence of donor cells. Transplanted GFP(+) donor cells (stained brown) proliferated and constituted the bone matrix of the specimens.

Additionally, because green florescence was not confirmed directly after making paraffin sections, anti-GFP immunostaining was performed to confirm the presence of donor cells. The bone matrix of specimens extirpated at 4 weeks after transplantation, was derived from the transplanted GFP(+) donor ASCs that had proliferated (Figure 14A, B). These results indicated that the ASCs on the hydroxyapatite surface in the micropores, were differentiated into osteoblasts. At this stage, we do not have incontrovertible proof that the ASCs differentiated into hematopoietic cells. However, this study demonstrated not only bone matrix regeneration but also bone marrow regeneration, as well as a reduction in GFP expression over time.

CONCLUSION

1. The Concept of adipose-derived Stem Cells

It was reported in 2001 that a multipotent cell population had been isolated from fluid drained during liposuction [2]; these were termed "processed lipoaspirate (PLA) cells". PLA cells are a pluripotent cell population in adipose tissue, and have been recognized as MSCs. Traditionally, MSCs referred only to bone marrow-derived cells [1]; however, since there are now alternative sources, it is necessary to state the origin of the MSCs used in each study [14]. At the 2nd annual meeting of the International Fat Applied Technology Society (IFATS, Pittsburgh, 2004), terminology was discussed and it was concluded that the term "ASCs" should be used for pluripotent cells harvested from adipose tissue. Thus, in this report, we describe adipose-derived stem mesenchymal stem cells as "ASCs" and bone marrow-derived mesenchymal stem cells as "BSCs."

As first reported in 2001 [2], ASCs have many attributes: they can be harvested, handled, and multiplied non-invasively, easily, and effectively; their pluripotency and proliferative efficiency are no less than those of BSCs; and donor morbidity is lower than for MSCs harvested from other sites. Mature adipocytes are particularly easy to remove from adipose tissue, requiring only collagenase treatment and centrifugation, since they are less dense than water. From a clinical point of view, sufficient quantities of MSCs could be harvested and cultured from a single gram of fat. Furthermore, only local anesthesia is required, and the wound at the donor site heals within a week. If the donor patient has liposuction or a lipectomy, a sufficient number of MSCs could be harvested for immediate auto transplantation, without culturing through *ex vivo* passages.

In 2002, Zuk et al. [7] reported that human ASCs expressed multiple CD marker antigens, similar to those observed on BSCs. They studied the cells as they differentiated not only into mesodermal lineages but also into neural lineages. ASCs differentiated into putative neural cells, exhibiting a neuronal-like morphology and expressing several proteins consistent with the neuronal phenotype. To confirm the multipotency of ASCs, comparisons between ASCs and BSCs have been made. De Ugarte et al. [15] showed that both human ASCs and BSCs express CD13, CD29, CD44, CD90, CD105, SH-3, and STRO-1. However, differences in expression were noted for the cell adhesion molecules CD49d (Integrin a4), CD54 (ICAM-1), CD34, and CD106 (VCAM-1). Thus, ASCs and BSCs have similar surface phenotypes, although they are distinct for several cell adhesion molecules that are implicated in hematopoietic stem cell homing, mobilization, and proliferation [15].

Unlike ES or EG cells, which are considered to be completely undifferentiated, our electron microscopic observation of mouse cells suggested that ASCs are not completely undifferentiated, as the cytoplasmic organelles are relatively developed. Moreover, when compared to our studies of BSCs, the nucleocytoplasmic ratio ($R_{N/C}$) was low. These results support the suggestion that ASCs harvested from mice have a greater degree of differentiation, than BSCs or ES cells [16]. However, we did confirm their pluripotency to differentiate into at least four different cell lineages, including a neural lineage. Furthermore, few chromatin granules were accumulated around the nuclei of ASCs, showing that these cells are in mitotic phase [17], and so it is believed that their proliferation potential is high. Thus, given their ability to differentiate and proliferate, ASCs are considered to be particularly suited to clinical applications.

In addition to human ASC studies, including that of Mizuno et al. [18], animal studies using rats or mice have been reported since 2002. In 2003, Dragoo et al. [19] performed both *in vitro* and *in vivo* studies with human ASCs. When they transformed MSCs with the bone morphogenic protein (BMP)-2 gene and implanted them into SCID mice, they were able to confirm bone formation *in vivo*. Thereafter, Morizono et al. [20] suggested the possible use of ASCs as gene delivery vehicles. In 2004, Cowan et al. [21] reported the treatment of critical-size mouse calvarial defects using ASCs and a scaffold made of apatite-coated polylactic-co-glycolic acid (PLGA). Around the same time, we started both *in vivo* and *in vitro* studies using ASCs, in particular GFP(+) ASCs [9,13].

2. Availability of GFP Transgenic Animals for Stem Cell Study

Green fluorescent protein (GFP), derived from the jellyfish *Aequorea victoria,* was first reported by Shimomura et al. [22] in 1962. The jellyfish yields a natural green fluorescence, which can be viewed in a dark environment [5]. GFP has recently become one of the most exciting and useful molecules in the field of biotechnology, particularly in the field of bioimaging. When expressed in cells and illuminated by UV light, GFP emits a bright green glow. Thus, we can detect both protein location and concentration, by the position and intensity of the glow, respectively.

GFP transgenic animals and plants such as *Drosophila melanogaster* [23], *Danio rerio* [24], and *Arabidopsis thaliana* [25], have been produced using wild-type GFP. In 1997, Okabe et al. produced transgenic mouse lines with an 'enhanced' GFP (EGFP), in which the cDNA was under the control of a chicken beta-actin promoter and cytomegalovirus enhancer [5]. All of the tissues from these transgenic lines, with the exception of erythrocytes and hair, were green under light excitation [5]. In 2001, Hakamata et al. [6] produced a EGFP transgenic rat.

The fluorescent nature of the cells from these transgenic animals provides an indication of the cell source for *in vivo* studies, and the cell population can be identified easily through fluorescence microscopy. Thus, we have studied stem cells using GFP transgenic mice and rats [9,13]. Of course, other methods, such as pre-labeling *in vitro*, can be used to label cells for transplantation studies. However, these nuclear or membrane labels have disadvantages, since the label is progressively diluted as the cells proliferate *in vivo*. As gene expression in GFP transgenic animals does not undergo dilution, it is considered the most suitable reporter for long-term transplantation studies. Initially, for this work, we sought to determine if ASCs harvested from GFP transgenic animals could also differentiate into osteogenic, chondrogenic, adipogenic and neural lineages *in vitro*. Then, as an *in vivo* study, we tried bone and bone marrow regeneration using ASCs harvested from GFP transgenic mice.

3. In Vitro Study of adipose-derived Stem Cells

In the present study, ASCs were harvested easily from the inguinal fat pads of both GFP transgenic mice and rats. The ASCs were fibroblastic in appearance and bright green under the fluorescence microscope. These cells were demonstrated to be highly proliferative in cell culture, and we were able to control reproducibly the differentiation of ASCs from GFP transgenic mice and rats, into osteogenic, chondrogenic, adipogenic and neural lineages.

GFP(+) ASCs grown in osteogenic media exhibited changes in cell structure. Seven days after osteogenic induction, we observed a white extracellular matrix. We were able to confirm osteogenic differentiation through specific staining and genetic study. One week after osteogenic induction, we performed both alkaline phosphatase and von Kossa staining; osteogenesis was confirmed through positive staining. Osteogenic cells were also identified by RT-PCR and cDNA microarray, and we confirmed the increased expression of several osteogenic-related genes, including alkaline phosphatase, osteopontin and osteocalcin. Owen et al. [26] have described the developmental sequence of bone, which consists of three phases: proliferation with matrix secretion, matrix maturation, and matrix mineralization. Osteopontin and alkaline phosphatase are early osteogenic markers, and are characterized in the matrix maturation phase [27]. Osteocalcin is a late bone marker, secreted only by osteoblasts, and it also signals terminal osteoblast differentiation [28]. Our study suggested that ASCs derived from GFP transgenic mice and rats could be differentiated into osteogenic lineages, and that after 4 weeks of incubation in osteogenic medium, ASCs still fluoresced green under fluorescence microscopy, although the number of cells decreased with an increase in the matrix.

In the chondrogenic differentiation study, the micromass culture technique reported previously [10], combined with media supplemented with transforming growth factor-beta 1 (TGF-beta 1), resulted in the condensation of the GFP(+) ASCs into three-dimensional aggregates. Alcian blue staining is specific for the highly sulfated proteoglycans of the cartilage matrix [7], and these aggregates demonstrated positive staining in addition to fluorescing bright green under fluorescence microscopy. Additionally, expression of chondrocyte-specific genes, including aggrecan and type II/X collagen, was detected. Aggrecan is a cartilage proteoglycan, type II collagen is a cartilage-specific marker [29], and type X collagen is a late expression marker of hypertrophic chondrocytes [29]. Taken together, the above results strongly suggest chondrogenesis.

GFP(+) ASCs incubated in adipogenic medium are gradually metamorphosed into multilocular fat cells. These cells are considered to be pre-adipocytes, since they stain positively with Oil Red O, and expression of the gene for the adipose-specific transcription factor PPAR-gamma2 can be confirmed. Small fat lipids in the multilocular pre-adipocytes fuse together and become very large. After two weeks, ASCs derived from both mice and rats, and incubated in adipogenic medium, became unilocular fat cells *in vitro*. They naturally detached from the bottom of the dishes, at which point they were considered to be mature adipocytes.

Our findings demonstrated that the ASCs isolated from GFP transgenic mice and rats possessed the ability to undergo morphologic and phenotypic changes under neural induction. As previously reported, adipose tissue may include neuronal fibers [30]. However, unlike brown adipose tissue, the adipocytes in white adipose tissue are not directly in contact with nerve terminals [31]. Zuk et al. [2] showed that white adipose tissues contained stem cells, adipocytes, vasculature, and matrix. It is unlikely, therefore, that there are preexisting neural cells that continue to grow in passaged cultures. Moreover, under electron microscopy, we showed that neurally-induced cells possessed microtubules in their soma and processes, providing strong evidence that the neurally-induced cells in the present study possessed neuronal structures. Within a few hours of neural induction, the ASCs changed from fibroblast-like in shape, to bipolar and multipolar. Immunostaining revealed that the bipolar cells expressed nestin, and that the multipolar cells expressed MAP 2 or neurofilament-70. Nestin is an intermediate filament protein that is thought to be expressed at high levels in neural stem cells, although its expression has also been observed in myogenic, endothelial, and hepatic cells [32]. MAP 2 and neurofilament-70 are markers for mature neurons [33]. The morphological changes from bipolar to multipolar cells may represent neuronal maturation. Immunochemical studies, using GFAP as a marker for astrocytes, demonstrated that there were far fewer glial cells than neurons among the induced cells. ASCs tended to differentiate into neurocytes rather than astrocytes, under the conditions used in the present study.

In summary, this *in vitro* study describes new animal models for the study of ASCs. The results suggest that ASCs from GFP transgenic mice and rats have osteogenic, chondrogenic, adipogenic, and neural differentiation potential *in vitro*. Since these cell populations can be identified easily through fluorescence microscopy, GFP transgenic mice may be an ideal source of ASCs and other stem cells for further experiments in tissue engineering and regenerative medicine.

4. In Vivo Study of adipose-derived Stem Cells

In our present study on bone regeneration in mice, we demonstrated that bone matrix can be regenerated subcutaneously from ASCs *in vivo*. We used GFP transgenic mice as a source of ASCs, and GFP fluorescence provided a means of distinguishing the transplanted cells from those of the recipient. Transplanted ASCs proliferated and differentiated into osteoblasts in micropores on the surface of hydroxyapatites. This study demonstrated not only bone matrix regeneration, but also bone marrow regeneration. We have not yet obtained incontrovertible proof that ASCs differentiated into hematopoietic cells. However, we were able to regenerate a subcutaneous microenvironment in which hematopoietic cells survived, and this result may lead to new treatments for fibrotic bone marrow diseases such as idiopathic myelofibrosis and osteopetrosis [14]. Bone marrow engineering has great potential in the field of hematopoietic disease therapy, and regeneration of the hematopoietic system using ASCs might be useful in the future for treating hematopoietic stem cell diseases such as leukemia, in addition to fibrotic bone marrow diseases. We plan to continue looking for suitable scaffolds and methods for bone and bone marrow regeneration using ASCs.

However, down-regulation of GFP expression was a problem in our study. GFP expression of transplanted cells was reduced by 6 weeks after the operation. Even so, Mothe et al. [34] reported that the GFP expression by cells derived from GFP transgenic animals is much more sustained than that of cells expressing GFP from transduced viral vectors. Thus, we believe that transplantation using cells harvested from GFP transgenic animals is the best method available to date. We have yet to investigate the reason for down-regulation of GFP expression, and to discover the means of overcoming this problem. At present, there are six hypotheses explaining loss of GFP: 1. Donor cell death (apoptosis); 2. Donor cell death (necrosis); 3. Altered donor cell translation; 4. Down-regulation of GFP expression; 5. Rejection of infected donor cells; and 6. Antigenicity of the GFP molecule. We consider the first three possibilities unlikely, since we observed developed bone and bone marrow formations on the specimens 10 weeks after transplantation, that were similar to the developments that were seen at 4 weeks. The fourth hypothesis may be difficult to test, since GFP is expressed throughout the whole life of GFP transgenic mice. Hypothesis 5 is also considered unlikely, since the possibility of infection is low. Thus, at this stage, the most likely explanation for reduced GFP expression is antigenicity to the GFP molecule. Despite this apparent shortcoming, it was demonstrated that GFP is a useful experimental system for at least the first 6 weeks following transplantation.

5. Future Prospects for adipose-derived Stem Cell Study

ASCs are readily accessible in large quantities and may be extracted with minimal morbidity, overcoming the risks that are associated with obtaining stem cells from other regions, such as bone marrow or the subventricular zone of the brain. ASCs also provide a renewable population of cells that can be easily expanded in culture media. Therefore, they provide significant potential benefits for clinical use. In 2004, Lendeckel et al. [35] reported the first clinical application of ASCs. Now, in addition to bone fractures and defects, myocardial infarction, cerebral infarction, and spinal cord injuries may now be considered as potential targets for clinical therapy with ASCs. In conclusion, ASCs show considerable potential for tissue engineering. However, before clinical application can be undertaken, it will be necessary to continue the studies using animal models, in particular, the use of ASCs from GFP transgenic animals.

ACKNOWLEDGEMENTS

I would like to thank Tomoharu Kiyosawa, Department of Plastic and Reconstructive Surgery, National Defense Medical College, for his valuable suggestions. I also would like to thank Masaru Okabe at the Genome Information Research Center, Osaka University, and Yoji Hakamata at the Center for Molecular Medicine, Jichi Medical School for providing beautiful photographs of transgenic animals. Moreover, this review could not have been written without the cooperation of my colleagues at Nippon Medical School, and I apologize to the many scientists whose work was not cited due to space limitations.

REFERENCES

[1] Friedenstein, A.J. (1976). Precursor cells of mechanocytes. *Int Rev Cytol, 47*, 327-59.

[2] Zuk, P.A., Zhu, M., Mizuno, H., Huang, J., Futrell, J.W., Katz, A.J., Benhaim, P., Lorenz, H.P. & Hedrick , M.H. (2001). Multilineage cells from human adipose tissue: implications for cell-based therapies. *Tissue Eng, 7*, 211-8.

[3] Young, H.E., Steele, T.A., Bray, R.A., Hudson, J., Floyd, J.A., Hawkins, K., Thomas, K., Austin, T., Edwards, C., Cuzzourt, J., Duenzl, M., Lucas, P.A. & Black, A.C. Jr. (2001). Human reserve pluripotent mesenchymal stem cells are present in the connective tissues of skeletal muscle and dermis derived from fetal, adult, and geriatric donors. *Anat Rec, 264*, 51-62.

[4] Erices, A., Conget, P. & Minguell, J.J. (2000). Mesenchymal progenitor cells in human umbilical cord blood. *Br Jhaematol, 109*, 235-42.

[5] Okabe, M., Ikawa, M., Kominami, K., Nakanishi, T., & Nishimune Y. (1997). 'Green mice' as a source of ubiquitous green cells. *FEBS Lett, 407*, 313-9.

[6] Hakamata, Y., Tahara, K., Uchida, H., Sakuma, Y., Nakamura, M., Kume, A., Murakami, T., Takahashi, M., Takahashi, R., Hirabayashi, M., Ueda, M., Miyoshi, I., Kasai, N. & Kobayashi, E. (2001). Green fluorescent protein-transgenic rat: a tool for organ transplantation research. *Biochem Biophys Res Commun, 286*, 779-85.

[7] Zuk, P.A., Zhu, M., Ashjian, P., De Ugarte, D.A., Huang, J.I., Mizuno, H., Alfonso, Z.C., Fraser, J.K., Benhaim, P., & Hedrick, M.H. (2002). Human adipose tissue is a source of multipotent stem cells. *Mol Biol Cell, 13*, 4279-95.

[8] Huang, J.I., Beanes, S.R., Zhu, M., Lorenz, H.P., Hedrick, M.H. & Benhaim, P. (2002). Rat extramedullary adipose tissue as a source of osteochondrogenic progenitor cells. *Plast Reconstr Surg, 109*, 1033-41.

[9] Ogawa, R., Mizuno, H., Watanabe, A., Migita, M., Shimada, T., & Hyakusoku, H. (2004). Osteogenic and chondrogenic differentiation by adipose-derived stem cells harvested from GFP transgenic mice. *Biochem Biophys Res Commun, 313*, 871-7.

[10] Reddi, A.H. (1982). Regulation of local differentiation of cartilage and bone by extracellular matrix: a cascade type mechanism. *Prog Clin Biol Res, 110*, 261-8.

[11] Ogawa, R., Mizuno, H., Watanabe, A., Migita, M., Hyakusoku, H., Shimada, T. (2004). Adipogenic differentiation by adipose-derived stem cells harvested from GFP transgenic mice-including relationship of sex differences. *Biochem Biophys Res Commun, 319*, 511-7.

[12] Ashjian, P.H., Elbarbary, A.S., Edmonds, B., De Ugarte, D., Zhu, M., Zuk, P.A., Lorenz, H.P., Benhaim, P., & Hedrick, M.H. (2003). In vitro differentiation of human processed lipoaspirate cells into early neural progenitors. *Plast Reconstr Surg, 111*, 1922-31.

[13] Fujimura, J., Ogawa, R., Mizuno, H., Fukunaga, Y., & Suzuki, H. (2005). Neural differentiation of adipose-derived stem cells isolated from GFP transgenic mice. *Biochem Biophys Res Commun, 333*, 116-21.

[14] Ogawa, R. The Importance of Adipose-Derived Stem Cells and Vascularized Tissue Regeneration in the Field of Tissue Transplantation. (2006). *Current Stem Cell Research and Therapy, 1*, (IN PRESS).

[15] De Ugarte, D.A., Alfonso, Z., Zuk, P.A., Elbarbary, A., Zhu, M., Ashjian, P., Benhaim, P., Hedrick, M.H., and Fraser, J.K. (2003). Differential expression of stem cell mobilization-associated molecules on multi-lineage cells from adipose tissue and bone marrow. *Immunol Lett, 89*, 267-70.

[16] Swanson, J.A., Lee, M. & Knapp, P.E. (1991). Cellular dimensions affecting the nucleocytoplasmic volume ratio. *J Cell Biol, 115*, 941-8.

[17] Setterfield, G., Sheinin, R., Dardick, I., Kiss, G. & Dubsky, M. (1978). Structure of interphase nuclei in relation to the cell cycle. Chromatin organization in mouse L cells temperature-sensitive for DNA replication. *J Cell Biol, 77*, 246-63.

[18] Mizuno, H., Zuk, P.A., Zhu, M., Lorenz, H.P., Benhaim, P. & Hedrick, M.H. (2002). Myogenic differentiation by human processed lipoaspirate cells. *Plast Reconstr Surg, 109*, 199-209.

[19] Dragoo, J.L., Choi, J.Y., Lieberman, J.R., Huang, J., Zuk, P.A., Zhang, J., Hedrick, M.H. & Benhaim, P. (2003). Bone induction by BMP-2 tranducted stem cells derived from human fat. *Orthop Res, 21*, 622-9.

[20] Morizono, K., Ugarte, D.A., Zhu, M., Zuk, P.A., Elbarbary, A., Ashjian, P., Benhaim, P., Chen, I.S.Y. & Hedrick, M.H. (2003). Multilineage cells from adipose tissue as gene delivery vehicles. *Human Gene Therapy, 12*, 59-66.

[21] Cowan, C.M., Shi, Y.Y., Aalami, O.O., Chou, Y.F., Mari, C., Thomas, R., Quarto, N., Contag, C.H., Wu, B., & Longaker, M.T. (2004). Adipose-derived adult stromal cells heal critical-size mouse calvarial defects. *Nat Biotechnol, 22*, 560-7.

[22] Shimomura, O., Johnson, F.H. & Saiga, Y. (1962). Extraction, purification and properties of aequorin, a bioluminescent protein from the luminous hydromedusan, Aequorea. *J Cell Comp Physiol, 59*, 223-39.

[23] Wang, S. & Hazelrigg, T. (1994). Implications for bcd mRNA localization from spatial distribution of exu protein in Drosophila oogenesis. *Nature, 369*, 400-03.

[24] Peters, K.G., Rao, P.S., Bell, B.S. & Kindman, L.A. (1995). Green fluorescent fusion proteins: powerful tools for monitoring protein expression in live zebrafish embryos. *Dev Biol, 171*, 252-7.

[25] Sheen, J., Hwang, S., Niwa, Y., Kobayashi, H. & Galbraith, D.W. (1995). Green-fluorescent protein as a new vital marker in plant cells. *Plant J, 8*, 777-84.

[26] Owen, T.A., Aronow, M., Shalhoub, V., Barone, L.M., Wilming, L., Tassinari, M.S., Kennedy, M.B., Pockwinse, S., Lian, J.B., and Stein, G.S. (1990). Progressive development of the rat osteoblast phenotype in vitro: reciprocal relationships in expression of genes associated with osteoblast proliferation and differentiation during formation of the bone extracellular matrix. *J Cell Physiol, 143*, 420-30.

[27] Rodan, G. A., Rodan, S.B. (1984). Expression of the osteoblast phenotype. *Bone Miner Res, 2*, 244

[28] Tracy, R.P., Andrianorivo, A., Riggs, B.L. & Mann, K.G. (1990). Comparison of polyclonal and monoclonal antibody-based immunoassays for osteocalcin: a study of sources of variation in assay results. *J. Bone Miner Res, 5*, 451-461.

[29] Yoo, J.U., Barthel, T.S., Nishimura, K., Solchaga, L., Caplan, A.I., Goldberg, V.M., & Johnstone, B. (1998). The chondrogenic potential of human bone-marrow-derived mesenchymal progenitor cells. *J. Bone Joint Surg. Am, 80*, 1745-57.

[30] Napolitano, L. & Fawcett, D.W. (1958). The fine structure of brown adipose tissue in the newborn mouse and rat. *J Biophys Biochem Cytol, 4*, 685-92.

[31] Johnson, P.R. & Greenwood, M.R.C. (1988). Innervation and vascularization of white adipose tissue, in: L. Weiss (Eds.), *Cell and Tissue Biology, A Textbook of Histology*, sixth ed., Urban & Schwarzenberg, Baltimore, pp. 193-5.

[32] Sanchez-Ramos, J., Song, S., Cardozo-Pelaez, F., Hazzi, C., Stedeford, T., Willing, A., Freeman, T.B., Saporta, S., Janssen, W., Patel, N., Cooper, D.R. & Sanberg, P.R. (2000). Adult bone marrow stromal cells differentiate into neural cells in vitro. *Exp Neurol, 164*, 247-56.

[33] Safford, K.M., Hicok, K.C., Safford, S.D., Halvorsen, Y.D., Wilkison, W.O., Gimble, J.M. & Rice, H.E. (2002). Neurogenic differentiation of murine and human adipose-derived stromal cells. *Biochem Biophys Res Commun, 294*, 371-9.

[34] Mothe, A.J., Kulbatski, I., van Bendegem, R.L., Lee, L., Kobayashi, E., Keating, A. & Tator, C.H. (2005). Analysis of Green Fluorescent Protein Expression in Transgenic Rats for Tracking Transplanted Neural Stem/Progenitor Cells. *J Histochem Cytochem, 27*.

[35] Lendeckel, S., Jodicke, A., Christophis, P., Heidinger, K., Wolff, J., Fraser, J.K., Hedrick, M.H., Berthold, L. & Howaldt, H.P. (2004). Autologous stem cells (adipose) and fibrin glue used to treat widespread traumatic calvarial defects: case report. *J Craniomaxillofac Surg, 32*, 370-3.

In: DNA Research Trends
Editor: Marius P. Young, pp. 125-142
ISBN: 978-1-60021-646-6

Chapter V

OVERVIEW OF LONGEVITY-ASSOCIATED MITOCHONDRIAL DNA 5178 C/A POLYMORPHISM AND A DISCUSSION OF ITS MODULATION OF THE EFFECTS OF HABITUAL SMOKING ON SERUM TOTAL AND LDL CHOLESTEROL LEVELS IN MIDDLE-AGED JAPANESE MEN

Akatsuki Kokaze[1,*], Mamoru Ishikawa[1,2], Naomi Matsunaga[1], Masao Yoshida[1], Ryuji Makita[3], Masao Satoh[4], Koji Teruya[5], Kanako Sekiguchi[1], Yumi Masuda[1], Matsuko Harada[1], Yoshiko Uchida[1] and Yutaka Takashima[1]

[a1]Department of Public Health, Kyorin University School of Medicine, 6-20-2 Shinkawa, Mitaka-shi, Tokyo 181-8611, Japan,
[b2]Mito Red Cross Hospital, 3-12-48 Sannomaru, Mito-shi, Ibaraki 310-0011, Japan;
[c3]Department of General Medicine, Kyorin University School of Medicine, 6-20-2 Shinkawa, Mitaka-shi, Tokyo 181-8611, Japan;
[d4]Department of Allergy and Immunology, Saitama Medical School, 38 Morohongo, Moroyama, Iruma-gun, Saitama 350-0495, Japan;
[e4]Department of Public Health, Kyorin University School of Health Sciences, 476 Miyashita-machi, Hachioji-shi, Tokyo 192-8508, Japan.

* Correspondence concerning this article should be addressed to Akatsuki Kokaze, Department of Public Health, Kyorin University School of Medicine, 6-20-2, Shinkawa, Mitaka-shi, Tokyo, 181-8611, Japan. Tel: +81(422)47-5511 Ext 3495; Fax: +81(422)76-0366; E-mail: dawn@kyorin-u.ac.jp.

ABSTRACT

Mitochondrial DNA 5178 cytosine/adenine (Mt5178 C/A) polymorphism, which is also known as NADH dehydrogenase subunit-2 237 leucine/methionine (ND2-237 Leu/Met) polymorphism, is reportedly associated with longevity in the Japanese population, and the Mt5178A genotype may possess antiatherogenic properties. Individuals with the Mt5178A genotype may be more resistant to atherogenic diseases, such as myocardial infarction and cerebrovascular diseases, than those with Mt5178C. Moreover, Mt5178 C/A polymorphism has been reported to be associated with blood pressure, intraocular pressure, serum protein fraction levels, and serum lipid levels in healthy Japanese individuals.

In order to investigate the interaction between Mt5178 C/A polymorphism and habitual smoking on serum lipid levels, we performed a cross-sectional study in which healthy Japanese male subjects were classified into two groups based on body mass index (BMI): BMI ≤ 22 (n = 107); and BMI > 22 (n = 214). Interaction between Mt5178 C/A polymorphism and cigarette smoking was observed with regard to serum triglyceride (TG) levels in BMI ≤ 22 subjects ($P = 0.026$), and with regard to serum total cholesterol (TC) levels and serum low-density lipoprotein cholesterol (LDLC) levels in BMI > 22 subjects ($P = 0.019$ and $P = 0.043$, respectively). In BMI > 22 subjects, TC levels were significantly lower in non-smokers with Mt5178A than in smokers with Mt5178A ($P = 0.015$), while LDLC levels were lower in non-smokers with Mt5178A than in those with Mt5178C or in smokers with Mt5178A ($P = 0.047$ and $P = 0.055$, respectively). Although interaction between Mt5178 C/A polymorphism and habitual smoking was not observed with regard to TG levels in BMI > 22 subjects ($P = 0.308$), TG levels were significantly higher in smokers with Mt5178A than in non-smokers with Mt5178A as well as in non-smokers with Mt5178C ($P = 0.011$ and $P = 0.014$, respectively). Taken together, these results suggest that habitual smoking may deprive Japanese Mt5178A (ND2-237Met) genotypic men having BMI of > 22 of their antiatherogenic advantage.

Keywords: Mitochondrial DNA 5178 cytosine/adenine polymorphism; NADH dehydrogenase; longevity; habitual smoking; serum total cholesterol levels; serum low-density lipoprotein cholesterol levels.

1. INTRODUCTION

1.1. Mitochondrial DNA 5178 C/A Polymorphism

Numerous studies on mitochondria have revealed their cell biology, molecular genetics, energy metabolism and mitochondria-related diseases, as well as their application to anthropology and forensic science [1]. Over the past few decades, a considerable number of studies have been conducted regarding the association of mitochondria with aging and longevity [2,3]. As a result of these studies, several mitochondrial DNA polymorphisms have been reported to be associated with longevity [4-9]. Mitochondrial DNA 5178 cytosine/adenine (Mt5178 C/A) polymorphism, which results in NADH dehydrogenase subunit-2 237 leucine/methionine (ND2-237 Leu/Met) polymorphism, is known to be one of these longevity-associated mitochondrial DNA polymorphisms.

Tanaka and his colleagues, who performed a comprehensive sequence analysis of the entire mitochondrial genome in centenarians, both obese and non-obese young adults, patients with Parkinson's or Alzheimer's disease and diabetic patients [10], first suggested that Mt5178 C/A (ND2-237 Leu/Met) polymorphism was associated with longevity in Japanese people [4,5]. They reported that the frequency of Mt5178A (ND2-237Met), representing mitochondrial DNA haplogroup D, was significantly higher in centenarians than in the general population. The Mt5178A genotype is generally rare worldwide [11,12], and is believed to be as high as 40-45% in the Japanese populace [4,13-21]. There have been a number of reports regarding the association between Mt5178 C/A polymorphism and adult-onset diseases [14,16,17,22], while individuals with Mt5178A appear to be resistant to atherosclerotic diseases. These observations on Mt5178 C/A polymorphism and its association with life expectancy may help to explain why life expectancy in Japan is one of the longest in the world.

Wang et al. investigated the association of longevity-associated Mt5178 C/A polymorphism with the occurrence and clinical features of type 2 diabetes [14]. The frequency of the Mt5178A genotype was lower in patients with type 2 diabetes than in controls. Moreover, patients with a maternal history of diabetes tended to carry the Mt5178C genotype more frequently when compared to those without a maternal history of diabetes and controls. Mean age at onset of diabetes was significantly lower in patients with Mt5178C than in those with Mt5178A, while insulin treatment tended to be more frequent in Mt5178C genotypic patients. Matsunaga et al. investigated the significance of Mt5178 C/A polymorphism in type 2 diabetic subjects [15]. Using ultrasonography, they assessed the association between the Mt5178 C/A genotypes and early atherosclerotic changes of the bilateral carotid arteries. Mean intima-media thickness in the bilateral arteries was significantly greater in the Mt5178A genotypic group than in the Mt5178C genotypic group, and the Mt5178 C/A polymorphism was significantly correlated with mean intima-media thickness and the presence of plaques. They concluded that the Mt5178A genotype exerted an antiatherogenic effect in type 2 diabetic patients.

Mukae et al. investigated the relationship between longevity-associated Mt5178 C/A polymorphism and acute myocardial infarction [16]. They reported that the frequency of the Mt5178C genotype was significantly higher in patients with acute myocardial infarction than in control subjects, and that this difference was more evident in younger patients. They concluded that Mt5178C may promote the occurrence of acute myocardial infarction, and that the Mt5178C genotype was possibly a novel coronary risk factor in Japanese. Takagi et al. also reported that the frequency of Mt5178A genotype was significantly higher in controls than in subjects with myocardial infarction after adjustment for age, gender, body mass index (BMI), smoking status, hypertension, diabetes, hypercholesterolemia and hyperuricemia [22].

Aging is one of the risk factors associated with cerebrovascular diseases. It is therefore epidemiologically and etiologically interesting to investigate the correlation between longevity-associated Mt5178 C/A polymorphism and cerebrovascular diseases. Okubo et al. investigated Mt5178 C/A polymorphism and angiotensin I-converting enzyme gene polymorphism in patients with cerebrovascular disorder and age-matched normal controls [17]. Although there was no significant difference in the frequencies of angiotensin I-converting enzyme genotype between the cerebrovascular disorder patients and controls, the frequency of Mt5178C was significantly higher in cerebrovascular disorder patients than in normal controls. This report thus suggests that the Mt5178C genotype is one of the genetic risk factors for cerebrovascular diseases in Japanese individuals.

Interestingly, Aoyama et al. conducted a cross-sectional analysis of Trp64Arg polymorphism in the β_3-adrenergic receptor gene (BAR-3) and Mt5178 C/A (ND2-237 Leu/Met) polymorphism in order to determine whether these polymorphisms are correlated with lifestyle-related phenotypes and nutrient intake [18]. They showed that body fat content was significantly higher in subjects with the BAR-3 variant genotype (Arg/Arg or Arg/Trp) and Mt5178A when compared to those with the BAR-3 normal genotype (Trp/Trp) and Mt5178A, and those with BAR-3 normal genotype and Mt5178C. Carbohydrate energy ratio was significantly higher and animal protein intake was significantly lower in subjects with the BAR-3 normal genotype and Mt5178A than in those with the other three genotypic combinations. Combinations of BAR-3 polymorphism and Mt5178 C/A polymorphism therefore appear to have an influence on eating behavior. Moreover, Kato et al. reported that Mt5178 C/A (ND2-237 Leu/Met) polymorphism is associated with bipolar disorder [19]; individuals with Mt5178C are reportedly more susceptible to bipolar disorder than those with Mt5178A.

The leading cause of death in Japan is cancer, and the question of whether Mt5178 C/A polymorphism is associated with cancer risk is of particular interest. Iwao et al. reported no association between Mt5178 C/A polymorphism and six types of cancer (esophageal cancer, stomach cancer, colorectal cancer, lung cancer, breast cancer and malignant lymphoma) in the Japanese population [20]. However, they also reported that the interaction between Mt5178 C/A genotype, age and cancer, was significant in lung cancer subjects aged $\geq$ 60 years. Therefore, further investigation is required into the association between longevity-associated Mt5178 C/A polymorphism and susceptibility to cancer.

As discussed above, Mt5178 C/A polymorphism is widely associated with the pathologies of common diseases. In fact, these observations prompted us to investigate whether Mt5178 C/A polymorphism may be associated with biophysical and biochemical factors in healthy middle-aged subjects. Therefore, with a focus on preventive medicine or successful aging, we conducted cross-sectional studies on the association of Mt5178 C/A polymorphism with blood pressure [23], intraocular pressure [24], serum protein fraction levels [25,26] and serum lipid levels [21,27] in primarily healthy subjects.

1.2. Mt5178 C/A Polymorphism and blood Pressure

We previously reported that Mt5178 C/A (ND2-237 Leu/Met) polymorphism may influence both diastolic blood pressure (DBP) in Japanese women and the blood-pressure-increasing effect of habitual drinking in Japanese men [23]. In healthy Japanese women, DBP was significantly higher in those with the Mt5178A (ND2-237Met) genotype (n = 41; mean DBP ± SE. = 70.2 ± 1.2 mmHg) than in those with the Mt5178C (ND2-237Leu) genotype (n = 50; 67.0 ± 1.0 mmHg) (P = 0.040). After adjusting for age, BMI, habitual drinking, and habitual smoking, this difference also remained present (P = 0.022). In healthy Japanese men, Mt5178 C/A polymorphism was not found to be associated with blood pressure. However, the results of thorough data analysis suggested that Mt5178 C/A polymorphism may be indirectly associated with systolic blood pressure (SBP) and DBP via drinking frequency. After adjusting for age, BMI, and habitual smoking, SBP was revealed to be significantly higher in daily drinkers with the Mt5178C genotype ($n = 81$; least-square mean SBP ± SE. = 128.3 ± 1.5 mmHg) than in occasional drinkers ($n = 43$; 120.3 ± 2.1 mmHg) and non-drinkers ($n = 69$; 118.7 ± 1.6 mmHg) with the Mt5178C genotype ($P = 0.002$ and $P < 0.001$, respectively). Moreover, DBP was observed to be significantly higher in daily drinkers with Mt5178C ($n = 81$; least-square mean DBP ± SE. = 75.1 ± 1.0 mmHg) than in occasional drinkers ($n = 43$; 69.9 ± 1.4 mmHg) and non-drinkers ($n = 69$; 70.1 ± 1.1 mmHg) with Mt5178C ($P = 0.002$ and $P < 0.001$, respectively). However, in men with Mt5178A, no significant differences in blood pressure among daily drinkers, occasional drinkers and non-drinkers were observed. Therefore, this may be a novel gene-environmental (lifestyle) interaction on blood pressure.

The mechanism of the interaction described above has proven to be difficult fully elucidate. NADH dehydrogenase, complex I in the respiratory chain, is involved in ethanol-related production of reactive oxygen species (ROS) [28]. Superoxide radicals in and around vascular endothelial cells play a critical role in the pathogenesis of hypertension [29]. Mitochondria reportedly have a biophysical influence on blood pressure [30-32]. Therefore, it has been postulated that biochemical differences in the function of the ND2-237Leu and ND2-237Met isoforms of NADH dehydrogenase are involved in the interaction of ND2-237 Leu/Met polymorphism with habitual drinking on blood pressure. However, such a hypothesis requires further investigation.

1.3. Mt5178 C/A Polymorphism and Intraocular Pressure

We also reported that Mt5178 C/A polymorphism may be associated with intraocular pressure (IOP) in middle-aged Japanese men [24]. Because age-related changes in IOP are associated with ethnicity [33-36] and/or atherosclerosis [37,38], it is interesting from a biophysical standpoint to analyze the association between longevity-associated Mt5178 C/A polymorphism and IOP. In one such study, 386 Japanese men, who had not previously undergone any medical treatment for hypertension, ocular hypertension or glaucoma, were selected as subjects. Mean IOP was significantly higher in men with Mt5178C ($n = 234$; mean IOP ± SE. = 13.3 ± 0.2 mmHg) than in those with Mt5178A ($n = 152$; 12.7 ± 0.2 mmHg) ($P = 0.037$). This difference in mean IOP between Mt5178 C/A genotypes remained evident after adjusting for age, BMI, blood pressure, habitual smoking and habitual drinking. In addition, interactions between Mt5178 C/A polymorphism and habitual smoking or habitual drinking with regard to IOP were observed. Multiple regression analysis revealed that habitual smoking was significantly associated with IOP in men with Mt5178A (P for trend = 0.020), while daily alcohol consumption was significantly associated with IOP in those with Mt5178C (P for trend = 0.021). Mitochondria are reportedly involved in the regulation of IOP [39-41]. Such reports demonstrate the need for further investigation to elucidate the mechanisms of association between Mt5178 C/A polymorphism and IOP.

1.4. Mt5178 C/A and Serum Protein Fraction Levels

Age-related changes in serum protein, albumin, and globulin levels have been reported [42-44], and thus the association between longevity-associated Mt5178 C/A polymorphism and serum protein fraction levels was examined. Alpha-1, alpha-2, and beta globulin proportions were significantly higher in Mt5178A genotypic women (n = 44) than in Mt5178C genotypic women (n = 55) (P = 0.002, P = 0.006, and P = 0.008, respectively) [25]. Moreover, alpha-1, alpha-2, and beta globulin levels were significantly higher in Mt5178A genotypic women than in Mt5178C genotypic women (P = 0.001, P = 0.002, and P = 0.018, respectively). These differences in serum globulin levels between the Mt5178 C/A genotypes were more evident in premenopausal women than in postmenopausal women. However, no such differences in serum protein fraction levels were observed in men. Tobacco smoke [45-47], even environmental tobacco smoke [48], influences serum protein fraction levels. We therefore investigated whether there was a difference in the effects of habitual smoking on serum protein fraction levels in Mt5178 C/A genotypic men [26]. In Mt5178C genotypic men, alpha-1 and alpha-2 globulin levels were significantly higher in smokers than in non-smokers ($P < 0.001$ and P = 0.002, respectively), and the influence of smoking on these globulin levels depended on daily cigarette consumption. However, in men with the Mt5178A genotype, no significant differences were observed in alpha-1 or alpha-2 globulin levels between smokers and non-smokers.

Levels of alpha-1 antitrypsin, alpha-1 acid glycoproteins, and alpha-2 macroglobulin are reported to be strongly associated with degree of atherosclerosis [49]. These findings suggest that atherosclerosis may be more severe in smokers with Mt5178C than in non-smokers with Mt5178C, and Mt5178A genotypic men may be resistant to the aggravating effects of smoking on atherosclerosis. However, another point needs to be considered. Alpha-1 and alpha-2 globulin levels were significantly higher in premenopausal women with Mt5178A than in those with Mt5178C. Both alpha-1 antitrypsin and alpha-1 acid glycoproteins are electrophoretically related to alpha-1 globulin. Hashimoto et al. reported that alpha-1 antitrypsin levels were lower after menopause and were increased by estrogen, while alpha-1 acid glycoproteins were higher after menopause and were decreased by estrogen [50]. Denko and Gabriel reported that alpha-1 antitrypsin exhibited a biphasic change during the aging process, while alpha-1 acid glycoproteins decreased linearly as age increased [51]. Therefore, further investigation is required in order to confirm the differences in serum protein levels between young and old individuals and those between men and women.

1.5. Mt5178 C/A Polymorphism and Serum Lipid Levels

We previously reported that Mt5178 C/A polymorphism may be associated with serum lipid levels in the Japanese population [21]. In healthy Japanese men, after adjusting for age and BMI, high-density lipoprotein cholesterol (HDLC) levels were significantly higher in those with the Mt5178A genotype (n = 148; mean HDLC ± SE. = 57.8 ± 1.2 mg/dl) than in those with the Mt5178C genotype (n = 214; 54.4 ± 1.0 mg/dl) (P = 0.026). However, this significant difference disappeared after adjusting for drinking frequency. We therefore investigated the interaction between Mt5178 C/A polymorphism and habitual drinking on serum lipid levels in 321 healthy middle-aged Japanese men [27]. Interaction between Mt5178 C/A polymorphism and daily drinking on serum triglyceride (TG) levels was observed (P = 0.019). Moreover, interaction between Mt5178 C/A polymorphism and cigarette consumption on serum TG levels was also observed (P = 0.022). Multiple regression analysis showed that in Mt5178A genotypic men, daily drinking decreased TG levels (P = 0.025) and that cigarette consumption increased TG levels ($P < 0.001$), while in Mt5178C genotypic men, the effects of daily drinking and cigarette consumption on TG levels were unclear.

In healthy Japanese women, TG levels were significantly lower in those with the Mt5178A genotype (n = 44; 84.5 ± 6.8 mg/dl) than in those with the Mt5178C genotype (n = 55; 108.1 ± 6.1 mg/dl) (P = 0.012) [21]. This difference in TG levels between the Mt5178 C/A genotypes was more evident in postmenopausal women than in premenopausal women.

1.6. Aim of this Study

As discussed above, the antiatherogenic effects of the Mt5178A genotype have been reported in healthy subjects [21,26] and in type 2 diabetic patients [15]. We have also reported interactions between Mt5178 C/A polymorphism and cigarette smoking on serum TG levels [27], IOP [24], serum protein fraction levels [26] and red blood cell counts [52] in Japanese men. Habitual smoking is a major risk factor for atherosclerosis [53], and is harmful to mitochondria [54]. Moreover, body mass index (BMI) levels appeared to strongly influence the relationship between cigarette smoking and serum lipid levels in a previous study [27]. Therefore, in order to investigate whether Mt5178 C/A (ND2-237 L/M) polymorphism modulates the effects of cigarette smoking on serum lipid levels, we performed a cross-sectional study in which Japanese male subjects were classified into two groups based on BMI: BMI ≤ 22; and BMI > 22.

2. Subjects and Methods

2.1. Subjects

A total of 602 volunteers were recruited from among individuals visiting the Mito Red Cross Hospital for regular medical check-ups. This study was conducted in accordance with the Declaration of Helsinki (Edinburgh 2000) and was approved by the Ethical Committee of Kyorin University School of Medicine. Written informed consent was obtained from all volunteers before participation. Among the volunteers, 327 men were enrolled in this study, and these men did not have any chronic diseases or take antihypertensive, antihyperlipideamic or antidiabetic medications. However, 6 individuals with unclear drinking frequency were subsequently excluded from analysis. Therefore, subjects of this study comprised 321 Japanese men (age, 53.0 ± 7.7 y; mean ± SD).

2.2. Clinical Characteristics of Subjects

Determination of blood chemical and physical data was conducted as described previously [21]. Briefly, venous blood was drawn after a minimum fasting period of 12 h. Serum total cholesterol levels (TC) were measured by Determiner L TC II (Kyowa Medex) [55], serum HDLC levels by Determiner L HDL-C (Kyowa Medex) [56], and serum TG levels by Determiner L TG II (Kyowa Medex) [57]. Serum low-density lipoprotein cholesterol (LDLC) levels were calculated according to Friedewald's formula [58]. Body mass index was defined as the ratio of subject weight (kg) to the square of subject height (m). A survey of drinking and smoking habits was performed by means of a questionnaire. Habitual drinking was classified based on drinking frequency (daily drinkers; occasional drinkers, which included those who drink several times per week; and non- or ex-drinkers, which include those who drink a few times per month). Habitual smoking was classified into non- or ex-smokers and current smokers.

2.3. Mt5178 C/A Genotyping

DNA was extracted from whole blood using the DNA Extractor WB kit (Wako Pure Chemical Industries, Japan). The forward and reverse oligonucleotide primers used to determine the Mt5178 C/A genotype were: 5'-CTTAGCATACTCCTCAATTACCC-3' and 5'-CTGAATTCTTCGATAATGGCCCA-3'. Polymerase chain reaction (PCR) was performed with 50 ng genomic DNA in buffer containing 1.5 mM $MgCl_2$, 1.25 mM dNTPs and 0.5 μl Amplitaq DNA polymerase (GeneAmp, Perkin Elmer, USA). After, initial denaturation at 94°C for 5 min, PCR was conducted through 40 cycles as follows: denaturation at 94°C for 30 s, annealing at 60°C for 60 s and polymerase extension at 72°C for 10 min. The PCR product was then digested with restriction enzyme *Alu*I (Nippon Gene, Japan) and electrophoresed on 1.5% agarose gels. Gels were stained with ethidium bromide, visualized under UV light, and photographed. The absence of an *Alu*I site was designated as Mt5178A, and the presence of this restriction site was designated as Mt5178C.

2.4. Statistical Analyses

For all analyses, subjects were divided into two groups based on BMI: "BMI ≤ 22" (n = 107); and "BMI > 22" (n = 214). The cut-off point (BMI = 22) was determined based on reports about optimal BMI [59]. Statistical analyses were performed using SAS statistical software, version 6.12 for Windows (SAS Institute, Inc., Cary, NC, 1997). In analysis of covariates (ANCOVA), the Mt5178 C/A genotypes (Mt5178C = 0, Mt5178A = 1), habitual smoking (non- or ex-smokers = 0, current smokers = 1), and habitual drinking (non-drinkers = 0, occasional-drinkers = 1, daily-drinkers = 2) were numerically coded. Age and BMI were included as covariates in the models. *P* values of < 0.05 were considered statistically significant.

3. Results

The frequency of the Mt5178A (ND2-237Met) genotype was 40.2% (43/107) in the BMI ≤ 22 group and 39.7% (85/214) in the BMI > 22 group. No significant differences in Mt5178A genotypic frequency were observed between the two BMI-classified groups (P = 0.936). No significant differences in baseline characteristics were observed between the Mt5178 C/A genotypes in either the BMI ≤ 22 or BMI > 22 group (Table 1).

Table 1. Characteristics and serum lipid levels by genotype in the BMI≤ 22 and BMI > 22 groups

	BMI ≤ 22			BMI > 22		
	Mt51878C (n = 64)	**mt5178A (n = 43)**	***P* value**	**Mt5178C (n = 129)**	**Mt5178A (n = 85)**	***P* value**
Age (y)	53.38 ± 1.04	52.95 ± 1.27	0.798	53.36 ± 0.65	52.22 ± 0.80	0.273
BMI (kg/m^2)	20.19 ± 0.17	20.58 ± 0.21	0.158	24.58 ± 0.17	24.84 ± 0.21	0.342
Total cholesterol (mg/dl)	203.23 ± 4.04	200.12 ± 4.93	0.626	203.16 ± 3.07	202.66 ± 3.78	0.918
HDL cholesterol (mg/dl)	60.28 ± 2.22	65.00 ± 2.71	0.181	51.67 ± 1.08	53.33 ± 1.33	0.333
LDL cholesterol (mg/dl)	120.59 ± 3.92	113.87 ± 4.78	0.279	124.13 ± 2.96	121.48 ± 3.65	0.574
Triglyceride (mg/dl)	111.83 ± 7.27	106.24 ± 8.87	0.627	136.85 ± 5.70	139.25 ± 7.02	0.791

Age, body mass index (BMI), total cholesterol levels, high density lipoprotein (HDL) cholesterol levels, low density lipoprotein (LDL) cholesterol levels, and triglyceride levels are given as mean ± SE. All P values show the significance of differences between Mt5178 C/A genotypes. LDL cholesterol levels were calculated according to the Friedewald formula [58].

ANCOVA revealed interactions in the BMI ≤ 22 group (Table 2) between Mt5178 C/A (ND2-237 Leu/Met) polymorphism and habitual smoking with regard to TG levels (P = 0.026). Interaction between Mt5178 C/A and habitual drinking was also observed with regard to TC and TG levels (P = 0.028 and P = 0.046, respectively). Habitual drinking was significantly associated with TC, HDLC and LDLC levels (P = 0.027, P = 0.002, and P < 0.001, respectively), and BMI was significantly associated with TC, LDLC, and TG levels (P = 0.049, P = 0.039, and P = 0.006, respectively).

ANCOVA revealed interactions in the BMI > 22 group (Table 3) between Mt5178 C/A polymorphism and habitual smoking with regard to TC and LDLC levels (P = 0.019 and P = 0.043, respectively). Moreover, habitual smoking was significantly associated with TG levels (P = 0.005), habitual drinking was only significantly associated with HDLC levels (P < 0.001), and BMI was significantly associated with HDLC and TG levels (P < 0.001 and P = 0.006, respectively).

Table 4 shows the least square mean levels of TC, HDLC, LDLC and TG levels after adjusting for age, BMI and habitual drinking according to Mt5178 C/A polymorphism and smoking activity in both the BMI ≤ 22 and BMI > 22 groups. In subjects with BMI of ≤ 22, no significant differences were observed in serum lipid levels between the Mt5178 C/A genotypes. Although the differences did not reach significance, TG levels were lower in smokers with Mt5178C than in non-smokers with Mt5178C (P = 0.124), and higher in smokers with Mt5178A than non-smokers with Mt5178C (P = 0.105). In subjects with BMI of > 22, TC levels were significantly higher in smokers with Mt5178A than in non-smokers with Mt5178A (P = 0.015). In addition, LDLC levels were significantly lower in non-smokers with Mt5178A than in those with Mt5178C (P = 0.047), while these levels were higher, but not significantly so, in smokers with Mt5178A when compared to non-smokers with Mt5178A (P = 0.055).

Table 2. Analysis of covariance for serum lipid levels in subjects BMI of ≤ 22

	Total cholesterol *F* value	HDL cholesterol *F* value	LDL cholesterol *F* value	Triglyceride *F* value
Mt5178 C/A	0.22	0.78	0.60	0.42
Age	0.02	0.15	0.53	0.87
BMI	3.96*	3.17	4.40*	7.81*
Habitual smoking	0.01	3.06	1.44	0.02
Habitual drinking	3.76*	6.79**	8.75***	1.09
Mt5178 C/A × habitual smoking	0.05	0.22	0.13	5.12*
Mt5178 C/A × habitual drinking	3.71*	0.40	1.27	3.18*

HDL: high density lipoprotein; LDL: low density lipoprotein; BMI: body mass index. *P < 0.05, **P < 0.005, ***P < 0.001. In ANCOVA, some independent variables were numerically coded: Mt5178 C/A genotypes (Mt5178C = 0, Mt5178A = 1), habitual drinking (non- or ex-drinkers = 0, occasional-drinkers = 1, daily drinkers = 2), and habitual smoking (non- or ex-smokers = 0, current smokers = 1). Mt5178 C/A × smoking habit represents interaction between Mt5178 C/A genotypes and habitual smoking. Mt5178 C/A × habitual drinking represents interaction between Mt5178 C/A genotypes and habitual drinking.

Table 3. Analysis of covariance for serum lipid levels in subjects with BMI of > 22

	Total cholesterol *F* value	HDL cholesterol *F* value	LDL cholesterol *F* value	Triglyceride *F* value
Mt5178 C/A	0.08	1.19	0.62	0.09
Age	0.31	0.30	0.17	0.00
BMI	2.49	15.27***	3.72	7.76*
Habitual smoking	1.00	2.86	0.30	8.06**
Habitual drinking	1.38	9.29***	2.10	1.01
Mt5178 C/A × habitual smoking	5.55*	0.01	4.14*	1.05
Mt5178 C/A × habitual drinking	1.00	2.30	1.60	1.27

HDL: high density lipoprotein.; LDL: low density lipoprotein.; BMI: body mass index. *P < 0.05, **P < 0.005, ***P < 0.001. In ANCOVA, some independent variables were numerically coded: Mt5178 C/A genotypes (Mt5178C = 0, Mt5178A = 1), habitual drinking (non- or ex-drinkers = 0, occasional-drinkers = 1, daily drinkers = 2), and habitual smoking (non- or ex-smokers = 0, current smokers = 1). Mt5178 C/A × smoking habit represents interaction between Mt5178 C/A genotypes and habitual smoking. Mt5178 C/A × habitual drinking represents interaction between Mt5178 C/A genotypes and habitual drinking.

Table 4. Serum lipid levels in nonsmokers and smokers by genotype

	Mt5178C		Mt5178A	
BMI ≤ 22	Nonsmokers (n = 34)	Smokers (n = 30)	Nonsmokers (n =21)	Smokers (n = 22)
Total cholesterol (mg/dl)	202.72 ± 5.37	202.32 ± 6.30	200.50 ± 6.77	202.84 ± 6.66
HDL cholesterol (mg/dl)	64.10 + 2.94	59.53 ± 3.45	69.19 ± 3.70	61.47 ± 3.65
LDL cholesterol (mg/dl)	114.99 ± 5.02	123.62 ± 5.89	112.98 ± 6.33	117.49 ± 6.24
Triglyceride (mg/dl)	118.13 ± 9.62	95.84 ± 11.28	91.68 ± 12.13	119.50 ± 11.94
BMI > 22	Nonsmokers (n = 71)	Smokers (n = 58)	Nonsmokers (n = 55)	Smokers (n = 30)
Total cholesterol (mg/dl)	205.17 ± 4.19	201.09 ± 4.63	192.92 ± 4.92	212.26 ± 6.53†
HDL cholesterol (mg/dl)	52.66 ± 1.39	49.81 ± 1.53	53.10 ± 1.63	50.51 ± 2.16
LDL cholesterol (mg/dl)	126.96 ± 4.03	121.97 ± 4.46	114.47 ± 4.72*	129.03 ± 6.28
Triglyceride (mg/dl)	128.27 ± 7.65	146.55 ± 8.45	126.72 ± 8.97	163.55 ± 11.92*,†

Total cholesterol, high density lipoprotein (HDL) cholesterol, low density lipoprotein (LDL) cholesterol, and triglyceride levels are given as least-square mean ± SE adjusted for age, body mass index and habitual drinking. *$P < 0.05$ vs nonsmokers with mt5178C; †$P < 0.05$ vs nonsmokers with mt5178A.

4. DISCUSSION

The present study demonstrates the interaction between Mt5178 C/A (ND2-237 Leu/Met) polymorphism and cigarette smoking on serum lipid levels in middle-aged Japanese men. Significant interactions between Mt5178 C/A polymorphism and habitual smoking with regard to TG levels were observed in "BMI ≤ 22" subjects (Table 2). In "BMI > 22" subjects, interactions between Mt5178 C/A polymorphism and habitual smoking with regard to TC and LDLC levels were observed (Table 3); TC and LDLC levels were significantly higher in smokers than in non-smokers among Mt5178A (ND2-237Met) genotypic men, while among Mt5178C (ND2-237Leu) genotypic men, neither of these differed significantly between smokers and non-smokers (Table 4).

The Mt5178A genotype has been reported to exert antiatherogenic effects [15,21,26]. Individuals with Mt5178C were found to be more susceptible to not only type 2 [14], but also type 1 diabetes [60]. Individuals with Mt5178A are reportedly more resistant to atherogenic diseases, such as myocardial infarction [16,22] and cerebrovascular disorders [17]. However, the present results and the fact that high levels of LDLC are epidemiologically known to be a risk factor for atherosclerosis [53] suggest that habitual smoking may counteract the antiatherogenic effects of Mt5178A (ND2-237Met).

Epidemiological approaches have demonstrated the relationship of BMI and smoking with serum lipid levels in Japanese men [61,62]. In middle-aged Japanese men, BMI is positively associated with LDLC and TG levels, and negatively associated with HDLC levels, while habitual smoking is positively associated with TG levels, and negatively associated with HDLC levels [61]. Our findings indicate the involvement of Mt5178 C/A (ND2-237 Leu/Met) polymorphism in these relationships. In addition, interactions between genetic polymorphism and cigarette smoking on serum lipid levels have recently been reported [63-65]. Interaction between lipoprotein lipase-*Hind*III polymorphism, apolipoprotein E polymorphism and habitual smoking (gene-gene-environmental interaction) was reported with regard to HDLC levels [64]. It therefore appears that for personalized prevention of lifestyle-related atherogenic diseases, consideration of individual genetic backgrounds may possibly be more effective than conventional approaches. Activity of cholesteryl ester transfer protein (CETP), which is associated with *Taq*IB polymorphism in the CETP gene, is positively associated with LDLC levels in the Japanese population [66], while polymorphism in the intestinal fatty acid binding protein is also reportedly associated with LDLC levels [67]. Apolipoprotein B gene polymorphisms [68] and peroxisome proliferator activated receptor alpha polymorphism [69] reportedly influence TC and LDLC levels. Gene-gene interactions between these polymorphisms and Mt5178 C/A (ND2-237 Leu/Met) polymorphisms with regard to serum lipid levels must thus be considered.

Differences in the interaction between Mt5178 C/A (ND2-237 Leu/Met) polymorphism and habitual smoking with regard to serum lipid profiles were observed between subjects having BMI of ≤ 22 and those having BMI of > 22. Changes in gene-environmental interactions on serum lipid levels with BMI have been reported [65]. Regrettably, we have no data that clarifies the reasons for these BMI-related differences. Control of body weight reportedly improves lipid profile [70]; however, whether body weight control and/or cessation of smoking improves TC or LDLC levels in current smokers with Mt5178A (ND2-237Met) warrants further longitudinal study.

Longevity-associated mitochondrial genotypic differences in the effects of BMI and habitual smoking on serum lipid levels may be the result of functional differences between the ND2-237 Leu/Met variants. This amino acid change may bring about physiological and/or biochemical changes in NADH dehydrogenase. Reactive oxygen species (ROS) are reportedly involved in atherogenesis [71] and aging [3], and the mitochondrial respiratory chain is an important source of ROS [72,73]. Complex I of the mitochondrial respiratory chain (NADH dehydrogenase) consists of over forty different subunits [74] and plays a key role in the production of ROS [73,75]. Cigarette smoking, which is known to cause increases in ROS [76], attenuates NADH dehydrogenase activity [77]. The mechanism by which ND2-237 Leu/Met polymorphism modulates the effects of cigarette smoking on intraocular pressure [24], serum protein fraction levels [26] and red blood cell counts [52] is also thought to result from physiological and biochemical differences between the ND2-237 Leu/Met variants. However, there is currently a lack of clear data and further investigation of these mechanisms is required.

A potential limitation of the present study was the evaluation of habitual drinking based on the frequency of alcohol consumption. The present study confirms the interaction between ND2-237 Leu/Met variants and habitual drinking with regard to TC and TG levels in subjects with BMI of ≤ 22. Although we also used this evaluation in previous reports [23,24,27], the question of whether the ND2-237 Leu/Met variants are sensitive to alcohol-induced effects on serum lipid levels in a dose-dependent manner warrants further investigation.

In conclusion, longevity-associated Mt5178 C/A (ND2-237 Leu/Met) polymorphism may influence the effects of habitual smoking on TG levels in Japanese men with BMI of ≤ 22, and on TC and LDLC levels in those with BMI of > 22. To the best of our knowledge, these are novel gene-environmental (lifestyle) interactions on TC and LDLC levels. Cigarette smoking and overweight may counteract the antiatherogenic effects of Mt5178A (ND2-237Met). These genetic epidemiological findings may contribute to the establishment of personalized prevention of lifestyle-related atherosclerotic diseases, which may be more effective than conventional prevention approaches.

ACKNOWLEDGEMENTS

This study was supported in part by Grants-in-Aid from the Ministry of Education, Culture, Sports, Science and Technology of Japan (No. 14570355) and the Chiyoda Mutual Life Foundation.

REFERENCES

[1] Scheffler, I. E. (2001) A century of mitochondrial research: achievements and perspectives. *Mitochndrion 1*: 3-31.

[2] Van Remmen, H. & Richardson, A. (2001) Oxidative damage to mitochondria and aging. *Exp. Gerontol. 36*: 957-968.

[3] Driver, C. (2003) Mitochondrial interventions in aging and longevity. In S. I. S. Rattan (Ed.) *Biology of aging and its modulation vol. 5: Modulating aging and longevity*, (pp. 205-217). Dordrecht, The Netherlands: Kluwer Academic Publishers.

[4] Tanaka, M., Gong, J. S., Zhang, J., Yoneda, M. & Yagi, K. (1998) Mitochondrial genotype associated with longevity. *Lancet 351*: 185-186.

[5] Tanaka, M., Gong, J. S., Zhang, J., Yamada, Y., Borgeld, H. J. & Yagi, K. (2000) Mitochondrial genotype associated with longevity and its inhibitory effect on mutagenesis. *Mech. Ageing Dev. 116*: 65-76.

[6] Ivanova, R., Lepage, V., Charron, D. & Schächter, F. (1998) Mitochondrial genotype associated with French Caucasian centenarians. *Gerontology 44*: 349.

[7] De Benedictis, G., Rose, G., Carrieri, G., De Luca, M., Falcone, E., Passarino, G., Bonafé, M., Monti, D., Baggio, G., Bertolini, S., Mari, D., Mattace, R. & Franceschi, C. (1999) Mitochondrial DNA inherited variants are associated with successful aging and longevity in humans. *FASEB J. 13*: 1532-1536.

[8] Ross, O. A., McCormack, R., Curran, M. D., Duguid, R. A., Barnett, Y. A., Rea, I. M. & Middleton, D. (2001) Mitochondrial DNA polymorphism: its role in longevity of the Irish population. *Exp. Gerontol. 36*: 1161-1178.

[9] Niemi, A. K., Hervonen, A., Hurme, M., Karhunen, P. J., Jylhä, M. & Majamaa, K. (2003) Mitochondrial DNA polymorphisms associated with longevity in a Finnish population. *Hum. Genet. 112*: 29-33.

[10] Tanaka, M., Takeyasu, T., Fuku, N., Guo, L. J. & Kurata, M. (2004) Mitochondrial genome single nucleotide polymorphism and their phenotypes in the Japanese. *Ann. N.Y. Acad. Sci. 1011*: 7-20.

[11] Cann, R. L., Stoneking, M. & Wilson, A. C. (1987) Mitochondrial DNA and human evolution. *Nature 325*: 31-36.

[12] Yao, Y. G., Kong, Q. G. & Zhang, Y., P. (2002) Mitochondrial DNA 5178A polymorphism and longevity. *Hum. Genet. 111*: 462-463.

[13] Shimokata, H., Yamada, Y., Nakagawa, M., Okubo, R., Saido, T., Funakoshi, A., Miyasaka, K., Ohta, S., Tsujimoto, G., Tanaka, M., Ando, F. & Niino, N. (2000) Distribution of geriatric disease-related genotypes in the National Institute for Longevity Science, Longevity Study of Aging (NILS-LSA). *J. Epidemiol. 10*: S46-S55.

[14] Wang, D., Taniyama, M., Suzuki, Y., Katagiri, T. & Ban, Y. (2001) Association of the mitochondrial DNA 5178 A/C polymorphism with maternal inheritance and onset of type 2 diabetes in Japanese patients. *Exp. Clin. Endocrinol. Diabetes 109*: 361-364.

[15] Matsunaga, H., Tanaka, Y., Tanaka, M., Gong, J.S., Zhang, J., Nomiyama, T., Ogawa, O., Ogihara, T., Yamada, Y., Yagi, K. & Kawamori, R. (2001) Antiatherogenic mitochondrial genotype in patients with type 2 diabetes. *Diabetes Care 24*: 500-503.

[16] Mukae, S., Aoki, S., Itoh, S., Sato, R., Nishio, K., Iwata, T. & Katagiri, T. (2003) Mitochondrial 5178A/C genotype is associated with acute myocardial infarction. *Circ. J. 67*: 16-20.

[17] Ohkubo, R., Nakagawa, M., Ikeda, K., Kodama, T., Arimura, K., Akiba, S., Saito, M., Ookatsu, Y., Atsuchi, Y., Yamano, Y. & Osame, M. (2002) Cerebrovascular disorders and genetic polymorphisms: mitochondrial DNA5178C is predominant in cerebrovascular disorders. *J. Neurol. Sci. 198*: 31-35.

[18] Aoyama, M., Shidoji, Y., Saimei, M., Tsunawake, N. & Ichinose, M. (2003) Phenotypic linkage between single-nucleotide polymorphisms of β_3-adrenergic receptor gene and NADH dehydrogenase subunit-2 gene, with special reference to eating behavior. *Biochem. Biophys. Res. Commun. 309*: 261-265.

[19] Kato, T., Kunugi, H., Nanko, S. & Kato, N. (2000) Association of bipolar disorder with the 5178 polymorphism in mitochondrial DNA. *Am. J. Med. Genet. 96*: 182-186.

[20] Iwao N, Iwao S, Kobayashi F, Tajima K, Tanaka M, Atsuta Y, Tamakoshi A & Hamajima N. (2003) No association of the mitochondrial genotype (Mt5178A/C) with six cancers in a Japanese population. *Asian Pacific J. Cancer Prev. 4*: 331-336.

[21] Kokaze, A., Ishikawa, M., Matsunaga, N., Yoshida, M., Sekine, Y., Teruya, K., Takeda, N., Sumiya, Y., Uchida, Y. & Takashima, Y. (2001) Association of the mitochondrial DNA 5178 A/C polymorphism with serum lipid levels in the Japanese population. *Hum. Genet. 109*: 521-525.

[22] Takagi, K., Yamada, Y., Gong, J. S., Sone, T., Yokota, M & Tanaka, M. (2004) Association of a 5178C→A (Leu237Met) polymorphism in the mitochondrial DNA with a low prevalence of myocardial infarction in Japanese individuals. *Atherosclerosis 175*: 281-286.

[23] Kokaze, A., Ishikawa, M., Matsunaga, N., Yoshida, M., Sekine, Y., Sekiguchi, K., Harada, M., Satoh, M., Teruya, K., Takeda, N., Fukazawa, S., Uchida, Y. & Takashima, Y. (2004) Longevity-associated mitochondrial DNA 5178 A/C polymorphism and blood pressure in the Japanese population. *J. Hum. Hypertens. 18*: 41-45.

[24] Kokaze, A., Yoshida, M., Ishikawa, M., Matsunaga, N., Makita, R., Satoh, M., Sekiguchi, K., Masuda, Y., Uchida, Y. & Takashima, Y. (2004) Longevity-associated mitochondrial DNA 5178 A/C polymorphism is associated with intraocular pressure in Japanese men. *Clin. Experiment. Ophthalmol. 32*: 131-136.

[25] Kokaze, A., Ishikawa, M., Matsunaga, N., Yoshida, M., Sekine, Y., Teruya, K., Takeda, N., Satoh, M., Sumiya, Y., Uchida, Y. & Takashima, Y. (2002) Association of the longevity-associated mitochondrial DNA 5178 A/C polymorphism with serum protein fraction levels in healthy Japanese women. *Exp. Gerontol. 37*: 931-936.

[26] Kokaze, A., Ishikawa, M., Matsunaga, N., Yoshida, M., Sekine, Y., Sekiguchi, K., Satoh, M., Harada, M., Teruya, K., Takeda, N., Uchida, Y. & Takashima, Y. (2003) Longevity-associated mitochondrial DNA 5178 A/C polymorphism influences effects of cigarette smoking on serum protein fraction levels in Japanese men. *Mech. Ageing Dev. 124*: 765-770.

[27] Kokaze, K., Ishikawa, M., Matsunaga, N., Yoshida, M., Sekine, Y., Sekiguchi, K., Satoh, M., Harada, M., Teruya, K., Takeda, N., Uchida, Y., Tsunoda, T. & Takashima, Y. (2003) Longevity-associated mitochondrial DNA 5178 A/C polymorphism modulates effects of daily drinking and cigarette consumption on serum triglyceride levels in middle-aged Japanese men. *Exp. Gerontol. 38*: 1071-1076.

[28] Bailey, S. M., Pietsh, E. C. & Cunningham, C. C. (1999) Ethanol stimulates the production of reactive oxygen species at mitochondrial complex I and III. *Free Radic. Biol. Med. 27*: 891-900.

[29] Inoue, M., Sato, E. F., Park, A. M., Nishikawa, M., Kasahara, E., Miyoshi, M., Ochi, A. & Utsumi, K. (2001) Cross-talk between NO and oxyradicals, a supersystem that regulates energy metabolism and survival of animals. *Free Radic. Res. 33*: 757-770.

[30] Atlante, A., Seccia, T. M., Pierro, P., Vulpis, V., Marra, E., Pirrelli, A. & Passarella, S. (1998) ATP synthesis and export in heart left ventricle mitochondria from spontaneously hypertensive rat. *Int. J. Mol. Med. 1*: 709-716.

[31] Osanai, T., Tanaka, M., Kamada, T., Takahashi, K., Okada, S., Sirato, K., Magota, K., Kodama, S. & Okumura, K. (2001) Mitochondrial coupling factor 6 as a potent endogenous vasoconstrictor. *J. Clin. Invest. 108*: 1023-1030.

[32] Iwai, N., Katsuya, T., Mannami, T., Higaki, J., Ogihara, T., Kokame, K., Ogata, J. & Baba, S. (2002) Association between SAH, an acyl-CoA synthetase gene, and hypertriglyceridemia, obesity, and hypertension. *Circulation 105*: 41-47.

[33] Shiose Y. (1984) The aging effect on intraocular pressure in an apparently normal population. *Arch. Ophthalmol. 102*: 883-887.

[34] Klein, B. E. K., Klein, R. & Linton, K. L. P. (1992) Intraocular pressure in an American community. The Beaver Dam Eye Study. *Invest. Ophthalmol. Vis. Sci. 33*: 2224-2228.

[35] Bonomi, L., Marchini, G., Marraffa, M., Bernardi, P., De Franco, I., Perfetti, S., Varotto, A. & Tenna, V. (1998) Prevalence of glaucoma and intraocular pressure distribution in a defined population. The Egna-Neumarkt Study. *Ophthalmology 105*: 209-215.

[36] Lee, J. S., Lee, S. H., Oum, B. S., Chung, J. S., Cho, B. M. & Hong, J. W. (2002) Relationship between intraocular pressure and systemic health parameters in a Korean population. *Clin. Experiment. Ophthalmol. 30*: 237-241.

[37] Hayreh, S. S., Bill, A. & Sperber, G. O. (1994) Effects of high intraocular pressure on the glucose metabolism in retina and optic nerve in old atherosclerotic monkeys. *Graefe's Arch. Clin. Exp. Ophthalmol. 232*: 745-752.

[38] Harris, A., Jonescu-Cuypers, C., Martin, B., Kagemann, L., Zalish, M. & Garzozi, H. J. (2001) Simultaneous management of blood flow and IOP in glaucoma. *Acta Ophthalmol. Scand. 79*: 336-341.

[39] Chen, H. B., Yamabayashi, S., Ou, B. & Tsukahara S. (1996) Morphological changes in rabbit ciliary epithelium and blood-aqueous barriers after intravitreal 10 (-5) M endothelin-1. *Exp. Eye Res. 62*: 605-612.

[40] Ueda, J., Wentz-Hunter, K. K., Cheng, E. L., Fukuchi, T., Abe, H. & Yue, B. Y. J. T. (2000) Ultrastructural localization of myocilin in human trabecular meshwork cells and tissues. *J. Histochem. Cytochem. 48*: 1321-1329.

[41] Wentz-Hunter, K. K., Ueda, J., Shimizu, N. & Yue, B. Y. J. T. (2002) Myocilin is associated with mitochondria in human trabeculat meshwork cells. *J. Cell Physiol. 190*: 46-53.

[42] Herrmann, F. R., Safran, C., Levkoff, S. E. & Minaker, K. L. (1992) Serum albumin level on admission as a predictor of death, length of stay, and readmission. *Arch. Int. Med. 152*: 125-130.

[43] Sahyoun, N. R., Jacques, P. F., Dallal, G. & Russell, R. M. (1996) Use of albumin as a predictor of mortality in community-dwelling and institutionalized elderly populations. *J. Clin. Epidemiol. 49*: 981-988.

[44] Sullivan, D. H. (2001) What do the serum proteins tell us about our elderly patients? *J. Gerontol. 56A*: M71-M74.

[45] Wolf, G. T., Chretien, P. B., Weiss, J. F., Edwards, B. K. & Spiegel, H. E. (1982) Effects of smoking and age on serum levels of immune reactive proteins. *Otolaryngol. Head Neck Surg. 90*: 319-326.

[46] Banedek, I. H., Blouin, R. A. & McNamara, P. J. (1984) Influence of smoking on serum protein composition and the protein binding of drugs. *J. Pharm. Pharmacol. 36*: 214-216.

[47] Lindberg, G., Råstam, L., Gullberg, B., Lundbald, A., Nilsson-Ehle, P. & Hanson, B. S. (1993) Serum concentrations of total sialic acid and sialoglycoproteins in relation to coronary heart disease risk markers. *Atherosclerosis 103*: 123-129.

[48] Shima, M. & Adachi, M. (1996) Effects of environmental tobacco smoke on serum levels of acute phase proteins in schoolchildren. *Prev. Med. 25*: 617-624.

[49] Mori, T., Sasaki, J., Kawaguchi, H., Handa, K., Takada, Y., Matsunaga, A., Kono, S. & Arakawa, K. (1995) Serum glycoproteins and severity of coronary atherosclerosis. *Am. Heart J. 129*: 234-238.

[50] Hashimoto, S., Miwa, M., Akasofu, K. & Nishida, E. (1991) Changes in 40 serum proteins of post-menopausal women. *Maturitas 13*: 12-33.

[51] Denko, C. W. & Gabriel, P. (1981) Age and sex related levels of albumin, ceruloplasmin, alpha 1 antitrypsin, alpha 1 acid glycoprotein, and transferrin. *Ann. Clin. Lab. Sci. 11*: 63-68.

[52] Kokaze, A., Ishikawa, M., Matunaga, N., Yoshida, M., Makita, R., Satoh, M., Teruya, K., Sekiguchi, K., Masuda, Y., Harada, M., Uchida, Y. & Takashima Y. Interaction between longevity-associated mitochondrial DNA 5178 C/A polymorphism and cigarette smoking on hematological parameters in Japanese men. *Arch. Gerontol. Geriatr.* in press.

[53] Tegos, T. J., Kalodiki, E., Sabeta, M. M. & Nicolaides, A. N. (2001) The genesis of atherosclerosis and risk factors: a review. *Angiology 52*: 89-98.

[54] Banzet, N., Francois, D. & Polla, B. S. (1999) Tobacco smoke induces mitochondrial depolarization along with cell death: effects of antioxidants. *Redox Rep. 4*: 229-236.

[55] Allain, C. C., Poon, L. S., Chan, C. S., Richmond, W. & Fu, P. C. (1974) Enzymatic determination of total serum cholesterol. *Clin. Chem. 20*: 470-475.

[56] Sugiuchi, H., Uji, Y., Okabe, H., Irie, T., Uekama, K., Kayahara, N. & Miyauchi, K. (1995) Direct measurement of high-density lipoprotein cholesterol in serum with polyethylene glycol-modified enzymes and sulfated alpha-cyclodextrin. *Clin. Chem. 41*: 717-723.

[57] Fossati, P. & Prencipe, L. (1982) Serum triglyceride determined colorimetrically with an enzyme that produces hydrogen peroxide. *Clin. Chem. 28*: 2077-2080.

[58] Friedewald, W. T., Levy, R. I. & Fredrickson, D. S. (1972) Estimation of the concentration of low-density lipoprotein cholesterol in plasma, without use of the preparative ultracentrifuge. *Clin. Chem. 18*: 499-502.

[59] Tokunaga, K., Matsuzawa, Y., Kotani, K., Keno, Y., Kobatake, T., Fujioka, S. & Tarui, S. (1991) Ideal body weight estimated from the body mass index with the lowest morbidity. *Int. J. Obes. 15*: 1-5.

[60] Uchigata, Y., Okada, T., Gong, J.S., Yamada, Y., Iwamoto, Y. & Tanaka, M. (2002) A mitochondrial genotype associated with the development of autoimmune-related type 1 diabetes. *Diabetes Care 25*: 2106.

[61] Nakanishi, N., Nakamura, K., Ichikawa, S., Suzuki, K. & Tatara, K. (1999) Relationship between lifestyle and serum lipid and lipoprotein levels in middle-aged Japanese men. *Eur. J. Epidemiol. 15*: 341-348.

[62] Umeda, T., Kono, S., Sakurai, Y., Shinchi, K., Imanishi, K., Nishikawa, H., Ogawa, S., Katsurada, M., Wakabayashi, K., Honjo, S. & Todoroki, I. (1998) Relationship of cigarette smoking, alcohol use, recreational exercise and obesity with serum lipid atherogenicity: a study of self-defense officials in Japan. *J. Epidemiol. 8*: 227-234.

[63] Malin, R., Loimaala, A., Nenonen, A., Mercuri, M., Vuori, I., Pasanen, M., Oja, P., Bond, G., Koivula, T., Lehtimaki, T. (2001) Relationship between high-density lipoprotein paraoxonase gene M/L55 polymorphism and carotid atherosclerosis differs in smoking and nonsmoking men. *Metabolism 50*: 1095-1101.

[64] Corella, D., Guillen, M., Saiz, C., Portoles, O., Sabater, A., Folch, J. & Ordovas, J. M. (2002) Associations of LPL and APOC3 gene polymorphisms on plasma lipids in a Mediterranean population: interaction with tobacco smoking and the APOE locus. *J. Lipid Res. 43*: 416-427.

[65] Lussier-Cacan, S., Bolduc, A., Xhignesse, M., Niyonsenga, T. & Sing, C. F. (2002) Impact of alcohol intake on measures of lipid metabolism depends on context defined by gender, body mass index, cigarette smoking, and apolipoprotein E genotype. *Arterioscler. Thromb. Vasc. Biol. 22*: 824-831.

[66] Ikewaki, K., Mabuchi, H., Teramoto, T., Yamada, N., Oikawa, S., Sasaki, J., Taketa, K., Saito, Y. & Japan CETP Study Group. (2003) Association of cholesteryl ester transfer protein activity and TaqIB polymorphism with lipoprotein variation in Japanese subjects. *Metabolism 52*: 1564-1570.

[67] Galluzzi, J. R., Cupples, L. A., Otvos, J. D., Wilson, P. W. F, Schaefer, E. J. & Ordovas, J. M. (2001) Association of the A/T54 polymorphism in the intestinal fatty acid binding protein with variations in plasma lipids in The Framingham Offspring Study. *Atherosclerosis 159*: 417-424.

[68] Bentzen, J., Jørgensen, T. & Fenger, M. (2002) The effect of six polymorphisms in the apolipoprotein B gene on parameters of lipid metabolism in a Danish population. *Clin. Genet. 61*: 126-134.

[69] Tai, E. S., Demissie, S., Cupples, L. A., Corella, D., Wilson, P. W., Schaefer, E. J. & Ordovas, J. M. (2002) Association between the PPARA L162V polymorphism and plasma lipid levels. The Framingham Offspring Study. *Arterioscler. Thromb. Vasc. Biol. 22*: 805-810.

[70] Riebe, R., Greene, G. W., Ruggiero, L., Stillwell, K. M., Blissmer, B., Nigg, D. R. & Caldwell, M. (2003) Evaluation of a healthy-lifestyle approach to weight management. *Prev. Med. 36*: 45-54.

[71] Hogg N. (1998) Free radicals in disease. *Semin. Reprod. Endocrinol. 16*: 241-248.

[72] Wallace, D. C. (1992) Mitochondrial genetics: a paradigm for aging and degenerative diseases? *Science 256*: 628-632.

[73] Lenaz, G., Bovina, C., Aurelio, M. D'., Fato, R., Formiggini, G., Genova, M. L., Giuliano, G., Pich, M. M., Paolucci, U., Castelli, G. P. & Ventura, B. (2002) Role of Mitochondria in oxidative stress and aging. *Ann. NY Acad. Sci. 959*: 199-213.

[74] Walker, J. E., Skehel, J. M. & Buchanan, S. K. (1995) Structural analysis of NADH: ubiquinone oxidoreductase from bovine heart mitochondria. *Methods Enzymol. 260*: 14-34.

[75] Barja, G. (1998) Mitochondrial free radical production and aging in mammals and birds. *Ann. NY Acad. Sci. 854*: 224-238.

[76] Traber, M. G., van der Vliet, A., Reznick, A. Z. & Cross, C. E. (2000) Tobacco-related diseases. Is there role for anitioxidant micronutrient supplementation? *Clin. Chest Med. 21*: 173-187.

[77] Smith, P. R., Cooper, J. M., Govan, G. G., Harding, A. E. & Schapira, A. H. (1993) Smoking and mitochondrial function: a model for environmental toxins. *Q. J. Med. 86*: 657-660.

In: DNA Research Trends
Editor: Marius P. Young, pp. 143-160
ISBN: 978-1-60021-646-6

Chapter VI

RESPONSES TO UV RADIATION IN THE EXTREMELY HALOPHILIC MODEL ARCHEON, *HALOBACTERIUM* SP. NRC-1

Shirley J McCready* and Ivan Boubriak
School of Life Sciences, Oxford Brookes University,
Gipsy Lane, Oxford, OX3 0BP, UK.

ABSTRACT

The major mutagenic and cytotoxic damage to cells caused by ultraviolet light is due to the induction of photoproducts in DNA, principally cyclobutane pyrimidine dimers (CPDs) and (6-4) photoproducts (6-4PP). Most organisms have developed multiple strategies for surviving UV irradiation. These include methods of protection from UV light, pathways for repair of UV-damaged DNA and tolerance mechanisms that allow cells to survive in the presence of unrepaired UV lesions. There is a wide variation in the sensitivities of different organisms to UV, reflecting the different combinations of repair and tolerance mechanisms employed. These mechanisms have been well studied in bacteria and eukaryotes, but are much less well characterised in the third domain of life, the archaea. A useful model archaeon for studies of the response to UV is the sun-tolerant, extremely halophilic, *Halobacterium* sp. NRC-1. This organism is amenable to genetic, bioinformatics and physiological analysis. Furthermore its genome was the first halophilic archaeal genome to be sequenced and well-designed microarrays are available, making it an ideal model archaeon for functional studies and studies of gene expression. A functional analysis of putative repair genes identified from the genome sequence has revealed that bacterial-type nucleotide excision repair, employing *uvrA*, *uvrB* and *uvrC* gene products, is the only dark repair mechanism in *Halobacterium*, in spite of the presence of homologs of eukaryotic excision repair genes such as *rad3* and *rad25*. In

* Correspondence concerning this article should be addressed to Shirley J McCready, School of Life Sciences, Oxford Brookes University, Gipsy Lane, Oxford, OX3 0BP, UK. tel: - 44 1865 483366; email: sjmccready@brookes.ac.uk.

natural sunlight photoreactivation is likely to be the most significant repair mechanism, however, and photoreactivation of cyclobutane dimers, but not (6-4) photoproducts, can be accomplished by the Phr2 photolyase. Microarray analysis of UV-induced transcripts has highlighted a response more similar to the eukaryotic than the bacterial response; repair genes are not up-regulated after UV irradiation and there is no evidence of a coordinated SOS response such as that found in bacteria. The most significantly up-regulated genes are *radA1*, the archaeal homolog of bacterial *recA* and eukaryotic *rad51*, and genes encoding subunits of ribonucleotide reductase. These results are discussed in relation to other published work on repair systems in archaea and the response to environmental UV in other microorganisms.

Keywords: archaea; *Halobacterium*; halophile; DNA repair; UV damage; UV-inducible; microarrays; photolyases.

Introduction

Living cells have been exposed to UV light in sunlight since life began some 3.8 billion years ago. UV irradiance at the early Earth's surface was much higher than it is today and comprised all three UV wavebands – short-wave UV-C (below 290 nm), mid-wave UV-B (290-320 nm) and long-wave UV-A (320-400 nm). The rise in atmospheric oxygen levels that accompanied the emergence of photosynthetic organisms led to the formation of the ozone layer and in today's atmosphere UV-C wavelengths, which are the most damaging, are absorbed by oxygen and ozone and never reach the Earth's surface. UV-B is the most harmful UV radiation that we are exposed to and, in recent times, it has become apparent that depletion of the stratospheric ozone layer is resulting in elevated levels of UV-B radiation reaching the Earth's surface [1]. This is especially significant because the shorter the UV-B wavelength the more DNA damage is done [2,3] and ozone depletion leads to a shift towards the lower, most damaging, wavelengths.

The major mutagenic and cytotoxic UV damage is the induction of photoproducts in DNA, principally cyclobutane pyrimidine dimers (CPDs), pyrimidine(6-4)pyrimidone photoproducts (6-4PP) and the Dewar isomer of (6-4) photoproducts [4,5]. The presence of photoproducts in the DNA impairs DNA replication and can result in mutation and cell death.

Fortunately, the shortest UV wavelengths are the least penetrating and UV-B cannot penetrate through more than a few layers of cells in multicellular organisms, or through more than 100 microns of soil or, at most, a few meters of water in aquatic environments [6,7]. Many organisms protect vulnerable cells by means of UV-absorbing pigments such as melanins in animals and fungi [8,9] and flavonoids in plants [10,11]. Plant cells are also protected by compounds in the cuticle [12] and dividing cells at apical meristems are shielded by leaves or bracts. Microorganisms in exposed habitats are especially vulnerable because they are single-celled. They may protect themselves by phototaxic responses - moving away from UV [13,14] – or by forming biofilms [15], and some synthesise UV-protective pigments or extracellular substances such as polysaccharides [16].

In addition to protection from UV light, most organisms have developed a variety of other strategies for enabling cells to survive UV radiation. These include several pathways for repairing UV-damaged DNA, arrest at cell-cycle checkpoints and tolerance mechanisms, such as replicational lesion bypass, that allow cells to survive and divide when UV lesions in the DNA remain unrepaired [17-20]. The relative importance of these repair and tolerance mechanisms varies considerably from one organism to another [21].

Because of this variation in responses to UV, even organisms that are exposed to high levels of sunlight in their natural environments show considerable variation in their UV sensitivities [22,23] (Figure 1). Most molecular studies of UV damage and repair have focussed on a few model organisms – principally yeasts, human and rodent cells, and *E.coli* as the model for bacteria - in spite of the fact that none of these model organisms necessarily encounters high UV levels in the wild. We are only just beginning to appreciate the range and relative importance of diverse responses to UV in other organisms, especially those which are exposed to sunlight in natural conditions.

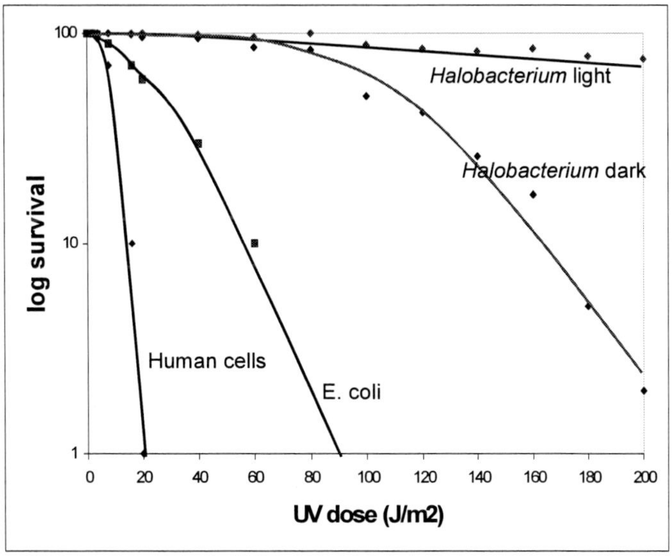

Figure 1. Relative sensitivities of selected organisms to UV radiation.

The Archaea

The archaea are a phylogenetically distinct third domain of life (Figure 2), distinct from bacteria and eukaryotes and they represent an important evolutionary link between eukaryotes and eubacteria. They were recognised as a third domain of living organisms, distinct from both prokaryotes and eukaryotes, in the late 1970s [24]. They resemble bacteria in their morphology and growth characteristics but the molecular biology of their replication, transcription and translation machinery resembles those of eukaryotes. The known Archaea are mainly 'extremophiles' living at extremes of temperature, pressure or high salt, although increasingly, archaea are identified in temperate environments [25-27]. It is becoming apparent that they are widely distributed and, in some environments they form a major part of the biomass [28].

Small subunit ribosomal RNA phylogeny distinguishes two main groupings of the archaea – the crenarchaeota and the euryarchaeota. The known crenarchaeota are mostly hyperthermophiles or thermoacidophiles, though they have also been found in mesophilic conditions. The euryarchaeota are more diverse and they have been found in a wider variety of environments. Many of these organisms are difficult to cultivate in the laboratory because of their requirement for extreme temperature or pressure. However, some are amenable to biochemical and molecular genetics analysis and there is now a large body of research on the archaea, their physiology, biochemistry, genomics and ecology and their potential for biotechnology.

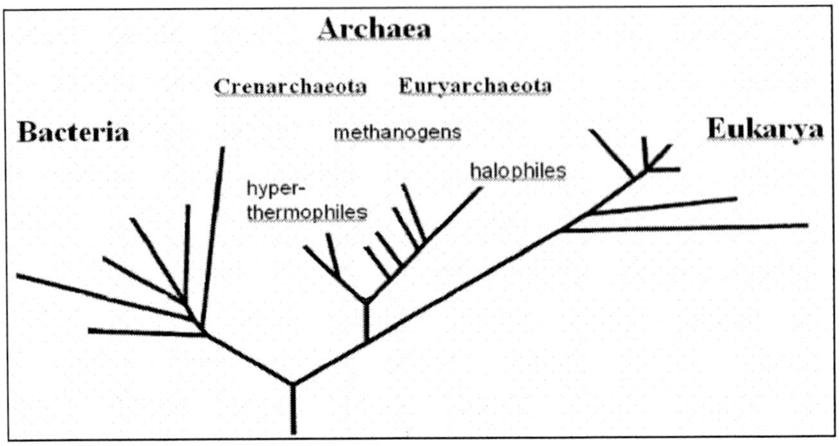

Figure 2. Outline of the three domains of life, based on relationships between small subunit ribosomal RNA sequences (originally analysed by Woese et al. [76]).

RESPONSE TO UV IN THE ARCHAEA

The archaea have an intriguing combination of prokaryotic and eukaryotic repair gene homologs. At the time of writing, 27 complete archaeal genome sequences have been published and these reveal a variety of putative repair genes [39-32]). There seem to be a number of different possible repair pathways represented but no obvious universal repair mechanism across the domain. Table 1 shows some of the putative repair gene homologs in archaeal genomes so far sequenced.

THE EXTREMELY HALOPHILIC ARCHAEA

The extreme halophiles are amongst the few archaeal groups that are exposed to strong sunlight in their natural environment (solar salt evaporation ponds and salt lakes, Figure 3) Several of the extreme halophiles have been used in laboratory studies, especially *Haloferax volcanii* and *Halobacterium salinarum*. At the time of writing, the complete genome sequences of *Halobacterium* sp. NRC-1 [29], *Haloarcula marismortui* ATCC 43049 [33] and *Natronomonas pharaonis* DSM 2160 [34] are available. Laboratory strains of *Halobacterium*

(Figure 4) are relatively UV resistant. *Halobacterium* is especially amenable to molecular genetic analysis and it is well proven as a model system for studying environmental stresses and for studies of gene expression [29,35-38]. The complete genome sequence of *Halobacterium* NRC-1, has been available since 2001 and tools for molecular genetic analysis are available – shuttle vectors, reporter genes, transformation, and microarray analysis. 36% of the genes in the *Halobacterium* genome have no known homologs so there are many novel genes to be characterised [29,39-41]. The response to UV irradiation of *Halobacterium* is clearly of interest because of the UV resistance of the organism and because of our relative ignorance about the archaea. An understanding of repair processes in the archaea can help us understand how repair and tolerance mechanisms evolved and how they relate to other cellular processes.

Figure 3. Extremely halophilic archaea in their natural hypersaline environment in salt evaporation ponds near El Jadida, Morocco [© Shirley McCready]

REPAIR OF UV DAMAGE IN *HALOBACTERIUM*

In common with other archaea, the genome sequence of *Halobacterium* reveals homologs of prokaryotic and eukaryotic repair genes [29] (Table 1).

Early work suggested that *Halobacterium salinarum* was unable to repair UV-induced DNA damage in the dark [42,43]. However, it was shown that *Halobacterium* is extremely efficiently photoreactivated, and survival is restored to near 100% after relatively high UV doses [42,44] so it was concluded that *Halobacterium* survived UV by photoreactivation alone. However, using a more sensitive assay for UV photoproducts, it was possible to show that *Halobacterium* does have a dark repair capability and that both cyclobutane pyrimidine dimers and (6-4) photoproducts are excised in the dark [45] (Figure 5). Both lesions are repaired more rapidly when exposed to visible light, which suggested that, in addition to the photolyase already known [46], there might be a (6-4) photoproduct photolyase.

Table 1. Repair gene homologues in selected sequenced archaeal genomes

	organism	Photo-reactivation	Alternative excision repair	Bacterial NER	Eukaryotic NER	Eukaryotic FLAP endonuclease
Crenarchaeota	*Sulfolobus tokadaii, S. solfataricus*	*phr*			XPB, XPF	FEN-1 (*rad2*)
	Sulfolobus acidocaldarius	*phr*	*uvsE*		XPB, XPF	FEN-1 (*rad2*)
	Pyrobaculum aerophilum				XPB, XPD, XPF	FEN-1 (*rad2*)
	Aeropyrum pernix				XPB, XPD, XPF	FEN-1 (*rad2*)
Euryarchaeota	*Thermoplasma volcanium, T. acidophilum*				XPB, XPD,	FEN-1 (*rad2*)
	Methanothermo-bacter thermo-autotrophicus	*phr*		*uvrA, uvrB, uvrC*	XPF	FEN-1 (*rad2*)
	Halobacterium sp NRC-1	*phr1, phr2*		*uvrA, uvrB, uvrC, uvrD*	XPB, XPD, XPF	FEN-1 (*rad2*)
extreme halophiles	*Haloarcula marismortui*	*phr1, phr2*	*uvsE*	*uvrA, uvrB, uvrC, uvrD*	XPB, XPD, XPF	FEN-1 (*rad2*)
	Natronomonas pharaonis	*phr1, phr2*		*uvrA, uvrB, uvrC, uvrD*	XPB, XPD, XPF	FEN-1 (*rad2*)
	Pyrococcus horikoshii, P. abyssi, P. furiosus				XPB, XPD, XPF	FEN-1 (*rad2*)
	Archaeoglobus fulgidus				XPB, XPD, XPF	FEN-1 (*rad2*)
	Methanopyrus kandleri	*phr*			XPB, XPF	FEN-1 (*rad2*)
	Methanococcus jannaschii				XPB, XPF	FEN-1 (*rad2*)
	Methanosarcina mazei, M. acetivorans, M. barkeri	*phr*		*uvrA, uvrB, uvrC, uvrD*	XPB, XPD, XPF	FEN-1 (*rad2*)

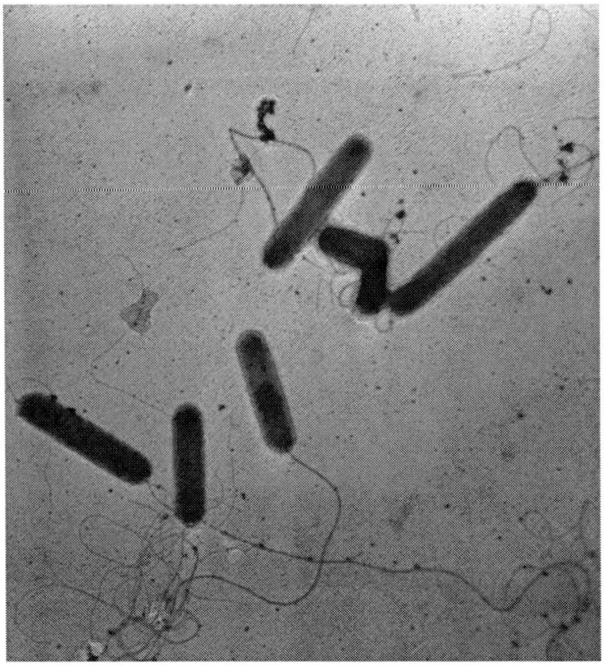

Figure 4. Electron micrograph of negatively stained *Halobacterium* sp. NRC-1. [© Shirley McCready]

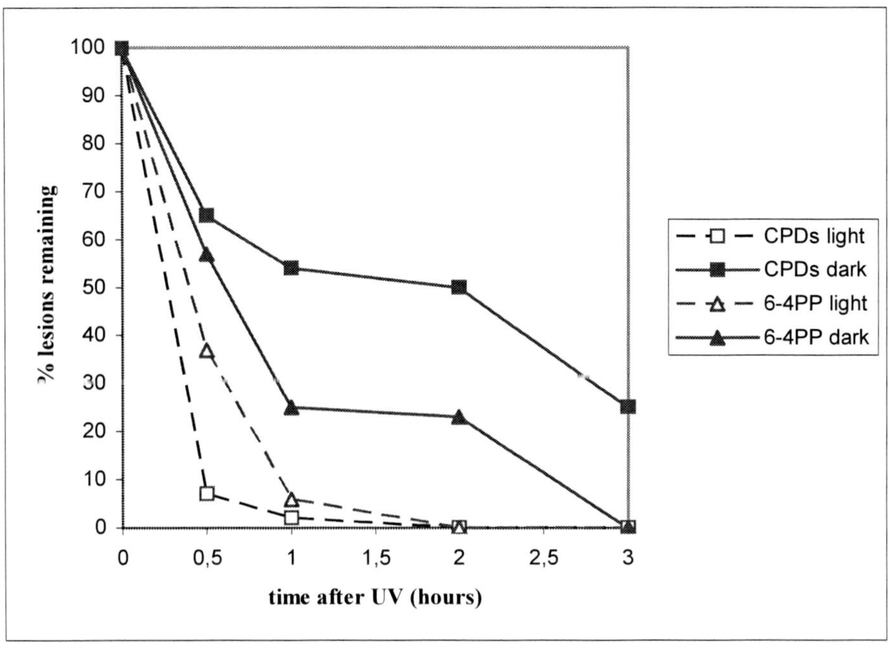

Figure 5. Repair of cyclobutane pyrimidie dimers (CPDs) and (6-4) photoproducts (6-4PP) in the dark and in the presence of photoreactivating light in wild-type *Halobacterium* sp. NRC-1 after a dose of 150 J/m^2 UV-C.

PHOTOREACTIVATION OF UV DAMAGE IN *HALOBACTERIUM*

Many organisms are able to repair UV photoproducts by the action of photolyases, enzymes that utilise photons of blue or near-UV light [47,48]. This is probably the principle repair pathway in organisms such as plants and aquatic microbes that are exposed to intense sunlight in their natural environment. Cyclobutane pyrimidine dimers make up the majority of UV photoproducts (70%-90%) and many prokaryotic and eukaryotic organisms have photolyases that repair CPDs [48,49]. A few organisms also have a (6-4) photoproduct photolyase, but these are much less common than CPD photolyases and no (6-4) photoproduct photolyase has been demonstrated in a prokaryote. Organisms that have (6-4) photoproduct photolyases include *Drosophila* [50] and *Arabidopsis* [51].

The *Halobacterium* genome contains two photolyase gene homologs, *phr1* and *phr2* [29]. Amongst the archaea, photolyases are not common and only the extreme halophiles are known to have two photolyase gene homologs. In *Halobacterium* sp. NRC-1 *phr1* and *phr2* are very closely related, with 38% amino acid sequence identity, so presumably they arose from a gene duplication. There is previous evidence that *phr2* encodes a CPD photolyase [46] and we have confirmed this by creating a deletion mutant lacking the *phr2* gene. This mutant was no more sensitive to UV light in the dark than in the light and its capacity to remove cyclobutane pyrimidine dimers from its DNA was the same in the dark as in the light [52].

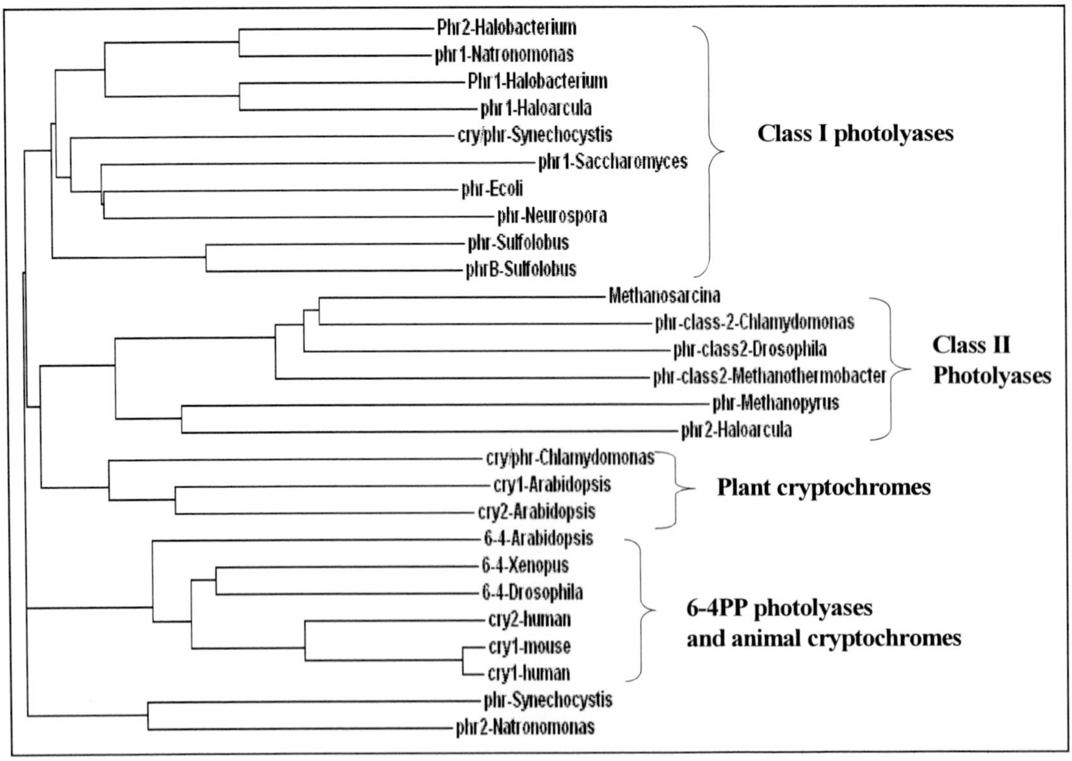

Figure 6. Cladogram of selected members of the photolyase/cryptochrome gene family (sequence alignment analysed using ClustalW).

Interestingly, the *phr2* deletion mutant still repaired 6-4PP more rapidly in the light than in the dark so it seemed possible that there may be a (6-4) photoproduct photolyase in *Halobacterium*. The obvious candidate for this would be *phr1*. To test this possibility, a *phr1* deletion mutant was created. However, the repair properties of the deletion mutant were not detectably different from the wild type, suggesting that *phr1* is not a (6-4) photoproduct photolyase (McCready and Boubriak, unpublished). This agrees with Baliga et al. who have shown no difference in UV sensitivity in a *phr1* deletion mutant compared to the wild type [53]. *Phr1* may be simply a pseudogene or, an alternative possibility is that it encodes a cryptochrome since photolyases and cryptochromes are closely related proteins [47] (Figure 6). The latter possibility is perhaps unlikely, however, since *phr1* is phylogenetically distant from cryptochromes in other organisms. The explanation of the apparent light-dependent repair of 6-4PPs may simply be that more energy is available for excision repair.

Excision Repair

Nucleotide excision repair (NER) is an almost universal repair mechanism used to repair CPDs and 6-4PPs induced by UV as well as bulky lesions resulting from other DNA-damaging agents. NER has been found in all bacteria and all eukaryotes. However, although the basic process is common to prokaryotes and eukaryotes – involving dual incisions either side of the UV photoproduct, removal of a single-stranded length of DNA containing the lesion, and resynthesis – the genes/proteins involved are different. In bacteria the key NER proteins are UvrA (damage recognition), UvrB and UvrC (which act as an exinuclease that cuts either side of the lesion), and UvrD (a DNA helicase). In *Saccharomyces cerevisiae*, the model for eukaryotes, the incisions are made by the Rad1/Rad10 complex on the 5‘ side and Rad2 (a member of the rad2/FEN-1 family of ‘flap’ endonucleases) on the 3‘ side of the UV photoproduct. Other proteins involved include Rad3 and Rad25 which are DNA helicases that work in opposite directions to open the repair bubble. These two proteins are subunits of the RNA polymerase II transcription initiation factor, TFIIH. A large number of other proteins are also involved in NER in eukaryotes [54].

There is also an alternative excision repair pathway (UVER) for repairing CPDs and 6-4PPs that is present in a number of eukaryotic and prokaryotic organisms, including fission yeast, the extremely radiation-resistant *Deinococcus radiodurans* and *Bacillus* species [21,55]. This mechanism is initiated by an endonuclease (Uvde/Uvd1/UvsE) which nicks the DNA immediately 5’ to a CPD or a 6-4PP. In *Schizosaccharomyces pombe* this is followed by either a further incision by Rad2, a structure-specific endonuclease homologous to the human FEN-1, and removal of the damage, or by endonuclease action of Exo1. Another repair mechanism for CPDs, but not 6-4PPs, that is present in some bacteria such as *Micrococcus luteus* [56] as well as bacteriophage T4 [57], is dependent on a UV endonuclease – DNA glycosylase/abasic lyase – which cleaves DNA at CPDs.

Excision Repair in the Archaea

DNA repair in the archaea is less well characterised and our knowledge is largely dependent on predicted gene functions from bioinformatics analysis of genome sequence information. Excision repair of UV damage has been demonstrated in *Halobacterium* NRC-1, *Sulfolobus solfataricus* and *Methanobacterium thermoautotrophicum*, [31,45,58,59] and in *M. thermoautotrophicum* the repair patch size has been shown to be 10-11 bases, characteristic of bacterial NER [59]. A few archaea – *Methanobacterium*, *Methanosarcina* and the halophiles *Haloarcula*, *Natronomonas* and *Halobacterium* – have homologs of the bacterial *uvrA*, *uvrB*, *uvrC* and *uvrD* NER genes [4,15,16]. The patch size observed in *Methanobacterium* suggests that the UvrABC excision repair pathway is probably functional in these archaea.

Interestingly, 'dark' repair has also been demonstrated in an archaeon that does not have *Uvr* gene homologs – *Sulfolobus solfataricus* [58]. UV-endonuclease-sensitive sites (CPDs) were observed to disappear within a few hours after irradiation with 200 J/m^2 of UV-C. In addition, transcripts of some of the eukaryotic repair gene homologs were shown, on Northern blots, to increase after UV, and it was suggested that *Sulfolobus* may have a 'stripped down' form of excision repair involving the products of the homologs of *rad1* (XPF) and *rad3* (XPD) although most of the other eukaryotic NER repair genes are absent in *Sulfolobus*, as in other archaea. Another possible explanation of this repair could be the presence of a UV endonuclease or a new, as yet uncharacterised repair system peculiar to the archaea [60].

Excision Repair in *Halobacterium* NRC-1

Halobacterium sp. NRC-1 is substantially more UV-resistant than fission yeast or *E. coli*, even in the absence of photoreactivating light. Given the presence of both bacterial and eukaryotic nucleotide excision repair genes identified in the genome sequence, the possibility of multiple excision repair pathways might seem a strong possibility, especially since excision repair was reported in *Sulfolobus* which lacks the bacterial *uvr* genes. Recently, however, a functional analysis has suggested otherwise. Deletion mutants lacking *uvrA*, *uvrB* and/or *uvrC* and have been shown to be unable to repair CPDs or 6-4PPs in the dark [31] (figure 7). This shows that only the bacterial type of excision repair is functional in *Halobacterium*.

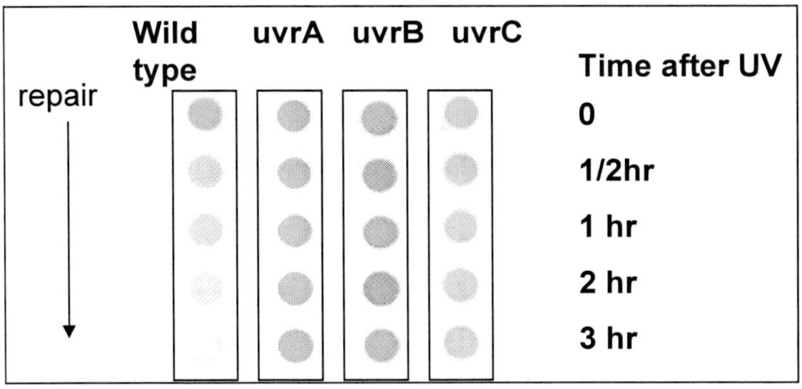

Figure 7. Immunoassay showing lack of repair in *Halobacterium* sp. NRC-1 deletion mutants lacking genes encoding proteins comprising the uvrABC repair pathway.

Inducible Responses to UV Irradiation

In addition to repair of UV damage, another important aspect of the response of cells to UV irradiation is the regulation of cell processes – at the levels of transcription, protein abundance and intracellular localisation. Expression of many genes in bacteria and eukaryotes has been shown to be induced by UV irradiation and different organisms have shown a variety of UV-inducible responses.

One of the principle responses to UV in bacteria is the inducible SOS response, in which LexA and RecA proteins mediate the regulation of over 40 genes, including excision repair genes (*uvrA*, *uvrB* and *uvrC*), *umuC, dinB* (encoding DNA polymerases that bypasses lesions), chaperonins and recombination genes. In addition, a number of genes are up-regulated independently of LexA [61], generally quite rapidly after UV and at lower levels of induction ($\leq$ 2-fold) than the LexA regulated genes.

DNA repair is not so obviously UV-inducible in eukaryotes and there is no coordinated SOS-like response. However, there are many genes that are up-regulated in response to UV. In simple eukaryotes these include genes encoding subunits of ribonucleotide reductase, bypass polymerases and some of the NER genes [54,62-64]. In budding yeast, *RAD9* controls induction of the DDR group of stress/damage-inducible genes. In human cells inducible genes include genes involved in cell cycle checkpoints, apoptosis and repair and inducibility by solar wavelengths has been studied as well as UV-C [65].

There have been few studies of UV-inducible genes in the archaea. *RadA*, the archaeal *rad51*/*recA* homolog, has been shown to be induced by UV in some archaea but apparently not in others [66,67]. Salerno et al. have published Northern blot analysis showing up-regulation of repair gene homologs in *Sulfolobus solfataricus* [58]. Baliga et al. have published microarray data showing transcript abundance in *Halobacterium* after UV irradiation [53]. However, they used an exceptionally high UV dose (200 J/m^2) and introduced other stresses such as changing media and chilling cells. A more recent microarray transcript analysis, using UV doses (30 J/m^2 and 70 J/m^2 UV-C) that were closer to biological doses and minimising stresses other than UV irradiation gave substantially different results [68]. (see: note on UV doses at the end of this chapter.)

This latter analysis revealed that, in *Halobacterium* in common with many other organisms, the most strongly upregulated genes are *radA1* (the *recA*/*rad51* homolog) and genes for subunts of ribonucleotide reductase. This pattern was seen in repeated experiments utilising 30 J/m^2 and also 70 J/m^2 UV-C, indicating the validity of the analysis. About 100 genes were up-regulated 1.5-fold or more in response to UV-C doses of 30 J/m^2 and 70 J/m^2. The response could be regarded as eukaryote-like. There was no evidence of a SOS response (in keeping with the lack of a *lexA* homolog in *Halobacterium*) [69]. UV repair genes were not up-regulated - neither the bacterial NER genes, *uvrA, uvrB, uvrC* and *uvrD* nor the homologs of eukaryotic *rad2*, *rad3*, *rad24* and *rad25* genes were up-regulated. Several genes involved in recombination were inducible: *RadA1* (*rad51*/*recA* homolog) was strongly up-regulated at all times and UV doses; vng2160 (*rpcA* replication protein A homolog), vng779 (predicted *recJ-like exonuclease*), *dbp* (eukaryote-like DNA-binding protein) and *hjr* (Holliday junction resolvase) were all up-regulated either at 1 hr or 3 hrs post-UV. Genes for ribonucleotide reductase subunits (*nrdB2*, *nrdB1* and *nrdA*) were strongly up-regulated as has been observed in many other organisms, presumably reflecting the need for rapid supply of nucleotides to allow accurate excision repair. At least a dozen up-regulated genes have no detectable similarity to any other genes in any organisms; a further 11 are unique to archaea and have unknown functions [68].

The Importance of RadA/RadA1

In the above transcriptional analysis *radA1* was highly up-regulated after both UV doses at 1hr and 3hr post-UV. The importance of r*adA* in halophilic archaea is indicated by the extreme UV-sensitivity of a *recA* deletion mutant of *Haloferax volcanii* [70]. In a number of archaea, the *radA*/*radA1* gene has been shown to be UV-inducible but the level of up-regulation of transcripts varies from one organism to another. In a study of several thermophilic archaea, using doses of UV-C between 10 J/m^2 and 50 J/m^2, the mesophiles showed a 6-fold up-regulation of *radA* transcript while up-regulation was only 2-fold in hyperthermophiles [66]. The micro-array data of McCready et al. is in keeping with this, showing 7-fold induction, similar to the mesophiles in this study [68].

Bacterial *recA* and its orthologs, archaeal *radA* and eukaryotic *RAD51*, are highly conserved. In bacteria, the RecA protein is induced by UV and has at least two functions in surviving UV damage – it is involved in regulation of the SOS response in which repair genes are up-regulated; and it is required for recombinational lesion bypass, which enables error-free DNA replication to continue past UV photoproducts [18,20]. RecA expression has been shown to be regulated in marine bacteria in response to natural sunlight [71].

In eukaryotes there is no SOS response and Rad51 has varying significance for UV survival in different organisms. *Rad51* mutants of *Saccharomyces cerevisiae* are only slightly UV sensitive but *Schizosaccharomyces pombe rhp51* mutants are very UV sensitive, suggesting that Rad51 is more important for UV survival in *S. pombe* – possibly because of the long G2 in *S. pombe* [72]. In many, but not all, eukaryotes *rad51* transcription is UV-inducible and in a few organisms the molecular mechanism of induction has been studied. The UV response in eukaryotic cells is complex and involves regulation of *rad51* expression as well as localisation of Rad51 protein to sites of DNA damage in the nucleus. Human *Rad51L* is UV-inducible, with maximal induction at 20 J/m^2 of UV-C [73]. Not all *RAD51* genes are up-regulated by UV, however – notably *S. cerevisiae RAD51* is not UV-inducible (though it is up-regulated by some other DNA-damaging agents including ionising radiation and aflatoxin [72]).

RadA1 is clearly important for survival of UV in *Halobacterium* even without the SOS response and further work will be directed towards its role in surviving UV irradiation and how its expression is controlled.

Conclusion

The sun-tolerant model archaeon, *Halobacterium* sp. NRC-1, has two repair pathways for UV damage – photoreactivation of CPDs, utilising the *phr2* photolyase, and bacterial-type excision repair utilising UvrA, UvrB and UvrC to remove CPDs and 6-4PPs. Compared to NER-deficient mutants in other organisms, such as *S. cerevisiae* and *E. coli*, *Halobacterium uvrA, uvrB* and *uvrC* deletion mutants are relatively resistant to UV [31] . This highlights the importance of other factors contributing to survival of UV irradiation. Studies of the transcriptional response to UV, using microarrays [68] have indicated the importance of the *radA1* gene in surviving UV irradiation in *Halobacterium*, in keeping with the observation in *Haloferax* that deletion of this gene results in UV sensitivity. The function of the radA1 protein is likely to be in stabilization of DNA replication forks stalled at UV photoproducts, enabling lesion bypass [18-20].

In addition to the repair and tolerance mechanisms discussed here, other factors likely to contribute to the relative resistance of *Halobacterium* to UV include its long cell division time (5 hours under favourable laboratory conditions), enabling time for repair prior to cell division. Another possible factor is the existence of multiple copies of the genome in *Halobacterium* cells [74].

There is clearly further work to be done to assess the relative contributions of these diverse factors as well as to understand in more detail the molecular basis for the response of *Halobacterium* to UV in sunlight.

Note on UV Doses

UV doses used in the laboratory for studies of repair and other responses to UV radiation are generally much higher than biological doses. The intensity of UV radiation varies greatly according to latitude, altitude and elevation of the sun. However, a number of published studies give measurements of the number of CPDs induced in DNA at various locations. One such study reports a maximum of 470 CPDs per megabase of DNA during 3 hours, in Antarctica, which equates to approximately 0.16 CPDs per kb per hour. We have also measured CPDs using a sensitive dot-blot technique and find the same order of magnitude –

i.e. 1 CPD per 7 kb per hour of sunlight which equates to 1.4 CPDs per kb per hour (McCready, unpublished data) in Western Canada at latitude 52° in August at 2 pm. This level of damage is induced by a dose of approximately 20 J/m^2 short-wave UV (UV-C) [75]. Thus, even the dose of 30 J/m^2 which we administered over a period of 30 seconds in our microarray experiments were higher than encountered in natural sunlight, so clearly higher doses up to 200 J/m^2 and above will give less indication of the natural response of organisms to UV.]

Acknowledgements

The author would like to thank the BBSRC for project grant # P18099 and Oxford Brookes University for financial support and Professor Shiladitya DasSarma and Dr David Crowley for their collaboration on the work reviewed in this chapter.

References

[1] Kerr JB, McElroy CT: Evidence for Large Upward Trends of Ultraviolet-B Radiation Linked to Ozone Depletion. *Science* 1993, 262:1033-1034.

[2] Setlow JK, Boling ME: Ultraviolet action spectra for mutation in Escherichia coli. *Mutat Res* 1970, 9(5):437-442.

[3] Coohill TP, Jacobson ED: Action spectra in mammalian cells exposed to ultraviolet radiation. *Photochem Photobiol* 1981, 33(6):941-945.

[4] Cadet J, Sage E, Douki T: Ultraviolet radiation-mediated damage to cellular DNA. *Mutat Res* 2005, 571(1-2):3-17.

[5] Mitchell DL, Nairn RS: The biology of the (6-4) photoproduct. *Photochem Photobiol* 1989, 49(6):805-819.

[6] Johnson D: Response of terrestrial microorganisms to ultraviolet-B radiation in ecosystems. *Res Microbiol* 2003, 154(5):315-320.

[7] Hader DP, Sinha RP: Solar ultraviolet radiation-induced DNA damage in aquatic organisms: potential environmental impact. *Mutat Res* 2005, 571(1-2):221-233.

[8] Johnson BE, Mandell G, Daniels F, Jr.: Melanin and cellular reactions to ultraviolet radiation. *Nat New Biol* 1972, 235(57):147-149.

[9] Zhdanova NN, Svishchuk AA, Shmygun MP, Bondar AI: [Protective effect of a pigment isolated from Cladosporium]. *Mikrobiologiia* 1970, 39(4):603-607.

[10] Solovchenko A, Schmitz-Eiberger M: Significance of skin flavonoids for UV-B-protection in apple fruits. *J Exp Bot* 2003, 54(389):1977-1984.

[11] Jordan BR, James PE, S AH-M: Factors affecting UV-B-induced changes in Arabidopsis thaliana L. gene expression: the role of development, protective pigments and the chloroplast signal. *Plant Cell Physiol* 1998, 39(7):769-778.

[12] Solovchenko A, Merzlyak M: Optical properties and contribution of cuticle to UV protection in plants: experiments with apple fruit. *Photochem Photobiol Sci* 2003, 2(8):861-866.

[13] Stoeckenius W: The rhodopsin-like pigments of halobacteria: light-energy and signal transducers in an archaebacterium. *Trends in Biochemical Sciences* 1985, 10:483-486.

[14] Ehling-Schulz M, Scherer S: UV protection in cyanobacteria. *European Journal of Phycology* 1999, 34:329-338.

[15] Elasri MO, Miller RV: Study of the response of a biofilm bacterial community to UV radiation. *Appl Environ Microbiol* 1999, 65(5):2025-2031.

[16] Ehling-Schulz M, Bilger W, Scherer S: UV-B-induced synthesis of photoprotective pigments and extracellular polysaccharides in the terrestrial cyanobacterium Nostoc commune. *J Bacteriol* 1997, 179(6):1940-1945.

[17] Lehmann AR: Clubbing together on clamps: The key to translesion synthesis. *DNA Repair (Amst)* 2006, 5(3):404-407.

[18] Kuzminov A: DNA replication meets genetic exchange: chromosomal damage and its repair by homologous recombination. *Proc Natl Acad Sci U S A* 2001, 98(15):8461-8468.

[19] Courcelle J, Hanawalt PC: RecA-dependent recovery of arrested DNA replication forks. *Annu Rev Genet* 2003, 37:611-646.

[20] Cox MM: Recombinational DNA repair of damaged replication forks in Escherichia coli: questions. *Annu Rev Genet* 2001, 35:53-82.

[21] Yasui A, McCready SJ: Alternative repair pathways for UV-induced DNA damage. *Bioessays* 1998, 20(4):291-297.

[22] Joux F, Jeffrey WH, Lebaron P, Mitchell DL: Marine bacterial isolates display diverse responses to UV-B radiation. *Appl Environ Microbiol* 1999, 65(9):3820-3827.

[23] Arrieta JM, Weinbauer MG, Herndl GJ: Interspecific variability in sensitivity to UV radiation and subsequent recovery in selected isolates of marine bacteria. *Appl Environ Microbiol* 2000, 66(4):1468-1473.

[24] Woese CR, Fox GE: Phylogenetic structure of the prokaryotic domain: the primary kingdoms. *Proc Natl Acad Sci U S A* 1977, 74(11):5088-5090.

[25] Abreu C, Jurgens G, De Marco P, Saano A, Bordalo AA: Crenarchaeota and Euryarchaeota in temperate estuarine sediments. *J Appl Microbiol* 2001, 90(5):713-718.

[26] Chaban B, Ng SY, Jarrell KF: Archaeal habitats from the extreme to the ordinary. *Can J Microbiol* 2006, 52(2):73-116.

[27] Simon HM, Jahn CE, Bergerud LT, Sliwinski MK, Weimer PJ, Willis DK, Goodman RM: Cultivation of mesophilic soil crenarchaeotes in enrichment cultures from plant roots. *Appl Environ Microbiol* 2005, 71(8):4751-4760.

[28] Karner MB, DeLong EF, Karl DM: Archaeal dominance in the mesopelagic zone of the Pacific Ocean. *Nature* 2001, 409(6819):507-510.

[29] Ng WV, Kennedy SP, Mahairas GG, Berquist B, Pan M, Shukla HD, Lasky SR, Baliga NS, Thorsson V, Sbrogna J *et al*: Genome sequence of Halobacterium species NRC-1. *Proc Natl Acad Sci U S A* 2000, 97(22):12176-12181.

[30] Grogan DW: The question of DNA repair in hyperthermophilic archaea. *Trends Microbiol* 2000, 8(4):180-185.

[31] Crowley DJ, Boubriak I, B. B, Clark M, Richard E, Sullivan L, S. D, McCready S: The uvrA, uvrB and uvrC genes are required for dark repair of ultraviolet light induced DNA photoproducts in Halobacterium sp. NRC-1. *Saline Systems* 2006, 2: 11

[32] DiRuggiero J, Brown JR, Bogert AP, Robb FT: DNA repair systems in archaea: mementos from the last universal common ancestor? *J Mol Evol* 1999, 49(4):474-484.

[33] Baliga NS, Bonneau R, Facciotti MT, Pan M, Glusman G, Deutsch EW, Shannon P, Chiu Y, Weng RS, Gan RR *et al*: Genome sequence of Haloarcula marismortui: a halophilic archaeon from the Dead Sea. *Genome Res* 2004, 14(11):2221-2234.

[34] Falb M, Pfeiffer F, Palm P, Rodewald K, Hickmann V, Tittor J, Oesterhelt D: Living with two extremes: conclusions from the genome sequence of Natronomonas pharaonis. *Genome Res* 2005, 15(10):1336-1343.

[35] Dassarma S, Berquist BR, Coker JA, Dassarma P, Muller JA: Post-genomics of the model haloarchaeon Halobacterium sp. NRC-1. *Saline Systems* 2006, 2:3.

[36] Dassarma S, Kennedy SP, Berquist B, Victor Ng W, Baliga NS, Spudich JL, Krebs MP, Eisen JA, Johnson CH, Hood L: Genomic perspective on the photobiology of Halobacterium species NRC-1, a phototrophic, phototactic, and UV-tolerant haloarchaeon. *Photosynth Res* 2001, 70(1):3-17.

[37] Soppa J: From genomes to function: haloarchaea as model organisms. *Microbiology* 2006, 152(Pt 3):585-590.

[38] Muller JA, DasSarma S: Genomic analysis of anaerobic respiration in the archaeon Halobacterium sp. strain NRC-1: dimethyl sulfoxide and trimethylamine N-oxide as terminal electron acceptors. *J Bacteriol* 2005, 187(5):1659-1667.

[39] Baliga NS, DasSarma S: Saturation mutagenesis of the TATA box and upstream activator sequence in the haloarchaeal bop gene promoter. *J Bacteriol* 1999, 181(8):2513-2518.

[40] Dyall-Smith M: The Halohandbook. Vers. 4.9: Copyright, Mike Dyall-Smith; 2004.

[41] Peck RF, Dassarma S, Krebs MP: Homologous gene knockout in the archaeon Halobacterium salinarum with ura3 as a counterselectable marker. *Mol Microbiol* 2000, 35(3):667-676.

[42] Sharma N, Hepburn D, Fitt PS: Photoreactivation in pigmented and non-pigmented extreme halophiles. *Biochim Biophys Acta* 1984, 799(2):135-142.

[43] Grey VL, Fitt PS: Evidence for the lack of deoxyribonucleic acid dark-repair in Halobacterium cutirubrum. *Biochem J* 1976, 156(3):569-575.

[44] Hescox MA, Carlberg DM: Photoreactivation in Halobacterium cutirubrum. *Can J Microbiol* 1972, 18(7):981-985.

[45] McCready S: The repair of ultraviolet light-induced DNA damage in the halophilic archaebacteria, Halobacterium cutirubrum, Halobacterium halobium and Haloferax volcanii. *Mutat Res* 1996, 364(1):25-32.

[46] Takao M, Kobayashi T, Oikawa A, Yasui A: Tandem arrangement of photolyase and superoxide dismutase genes in Halobacterium halobium. *J Bacteriol* 1989, 171(11):6323-6329.

[47] Deisenhofer J: DNA photolyases and cryptochromes. *Mutat Res* 2000, 460(3-4):143-149.

[48] Kanai S, Kikuno R, Toh H, Ryo H, Todo T: Molecular evolution of the photolyase-blue-light photoreceptor family. *J Mol Evol* 1997, 45(5):535-548.
[49] Yasui A, Eker AP, Yasuhira S, Yajima H, Kobayashi T, Takao M, Oikawa A: A new class of DNA photolyases present in various organisms including aplacental mammals. *Embo J* 1994, 13(24):6143-6151.
[50] Todo T, Takemori H, Ryo H, Ihara M, Matsunaga T, Nikaido O, Sato K, Nomura T: A new photoreactivating enzyme that specifically repairs ultraviolet light-induced (6-4) photoproducts. *Nature* 1993, 361(6410):371-374.
[51] Jiang CZ, Yee J, Mitchell DL, Britt AB: Photorepair mutants of Arabidopsis. *Proc Natl Acad Sci U S A* 1997, 94(14):7441-7445.
[52] McCready S, Marcello L: Repair of UV damage in Halobacterium salinarum. *Biochem Soc Trans* 2003, 31(Pt 3):694-698.
[53] Baliga NS, Bjork SJ, Bonneau R, Pan M, Iloanusi C, Kottemann MC, Hood L, DiRuggiero J: Systems level insights into the stress response to UV radiation in the halophilic archaeon Halobacterium NRC-1. *Genome Res* 2004, 14(6):1025-1035.
[54] Friedberg EC, Walker GC, Siede W: DNA Repair and Mutagenesis: Am. Soc. Microbiol., Washington, DC; 1995.
[55] Yonemasu R, McCready SJ, Murray JM, Osman F, Takao M, Yamamoto K, Lehmann AR, Yasui A: Characterization of the alternative excision repair pathway of UV-damaged DNA in Schizosaccharomyces pombe. *Nucleic Acids Res* 1997, 25(8):1553-1558.
[56] Shiota S, Nakayama H: UV endonuclease of Micrococcus luteus, a cyclobutane pyrimidine dimer-DNA glycosylase/abasic lyase: cloning and characterization of the gene. *Proc Natl Acad Sci U S A* 1997, 94(2):593-598.
[57] Lloyd RS, Hanawalt PC, Dodson ML: Processive action of T4 endonuclease V on ultraviolet-irradiated DNA. *Nucleic Acids Res* 1980, 8(21):5113-5127.
[58] Salerno V, Napoli A, White MF, Rossi M, Ciaramella M: Transcriptional response to DNA damage in the archaeon Sulfolobus solfataricus. *Nucleic Acids Res* 2003, 31(21):6127-6138.
[59] Ogrunc M, Becker DF, Ragsdale SW, Sancar A: Nucleotide excision repair in the third kingdom. *J Bacteriol* 1998, 180(21):5796-5798.
[60] Makarova KS, Aravind L, Grishin NV, Rogozin IB, Koonin EV: A DNA repair system specific for thermophilic Archaea and bacteria predicted by genomic context analysis. *Nucleic Acids Res* 2002, 30(2):482-496.
[61] Courcelle J, Khodursky A, Peter B, Brown PO, Hanawalt PC: Comparative gene expression profiles following UV exposure in wild-type and SOS-deficient Escherichia coli. *Genetics* 2001, 158(1):41-64.
[62] Yu S, Teng Y, Lowndes NF, Waters R: RAD9, RAD24, RAD16 and RAD26 are required for the inducible nucleotide excision repair of UV-induced cyclobutane pyrimidine dimers from the transcribed and non-transcribed regions of the Saccharomyces cerevisiae MFA2 gene. *Mutat Res* 2001, 485(3):229-236.
[63] Ruby SW, Szostak JW: Specific Saccharomyces cerevisiae genes are expressed in response to DNA-damaging agents. *Mol Cell Biol* 1985, 5(1):75-84.

[64] Lee JK, Park EJ, Chung HK, Hong SH, Joe CO, Park SD: Isolation of UV-inducible transcripts from Schizosaccharomyces pombe. *Biochem Biophys Res Commun* 1994, 202(2):1113-1119.

[65] Marrot L, Belaidi JP, Jones C, Perez P, Meunier JR: Molecular responses to stress induced in normal human caucasian melanocytes in culture by exposure to simulated solar UV. *Photochem Photobiol* 2005, 81(2):367-375.

[66] Reich CI, McNeil LK, Brace JL, Brucker JK, Olsen GJ: Archaeal RecA homologues: different response to DNA-damaging agents in mesophilic and thermophilic Archaea. *Extremophiles* 2001, 5(4):265-275.

[67] Sandler SJ, Satin LH, Samra HS, Clark AJ: recA-like genes from three archaean species with putative protein products similar to Rad51 and Dmc1 proteins of the yeast Saccharomyces cerevisiae. *Nucleic Acids Res* 1996, 24(11):2125-2132.

[68] McCready S, Muller JA, Boubriak I, Berquist BR, Ng WL, Dassarma S: UV irradiation induces homologous recombination genes in the model archaeon, Halobacterium sp. NRC-1. *Saline Systems* 2005, 1:3.

[69] Berquist B, al. e: Comparative genomic survey of information transfer systems in two diverse extremely halophilic Archaea... In: *Adaptation to Life at High Salt Concentrations in Archaea, Bacterial, and Eukarya.* Edited by Gunde-Cimerman N: Springer; 2005: 148-182.

[70] Woods WG, Dyall-Smith ML: Construction and analysis of a recombination-deficient (radA) mutant of Haloferax volcanii. *Mol Microbiol* 1997, 23(4):791-797.

[71] Booth MG, al. e: Quantification of RecA gene expression as an indicator of repair potential in marine bacterioplankton communities in Antarctica. *Aquatic Microbial Ecology* 2001, 24:51-59.

[72] Keller-Seitz MU, Certa U, Sengstag C, Wurgler FE, Sun M, Fasullo M: Transcriptional response of yeast to aflatoxin B1: recombinational repair involving RAD51 and RAD1. *Mol Biol Cell* 2004, 15(9):4321-4336.

[73] Peng L, Rice MC, Kmiec EB: Analysis of the human RAD51L1 promoter region and its activation by UV light. *Genomics* 1998, 54(3):529-541.

[74] Breuert S, Allers T, Gabi S, Soppa J: Regulated polyploidy in halophilic archaea. *PloS ONE* 2006, 1:e92.

[75] McCready SJ, Boyce JM, Cox BS: Excision repair in the yeast, Saccharomyces cerevisiae. *J Cell Sci Suppl* 1987, 6:25-38.

[76] Woese CR, Kandler O, Wheelis ML: Towards a natural system of organisms: proposal for the domains Archaea, Bacteria, and Eucarya. *Proc Natl Acad Sci U S A* 1990, 87(12):4576-4579.

In: DNA Research Trends
Editor: Marius P. Young, pp. 161-176
ISBN: 978-1-60021-646-6

Chapter VII

PARP-TYPE FINGERS AND SAP DOMAINS: STRUCTURE-SPECIFIC DNA RECOGNITION AND DNA REPAIR

Stefania Petrucco* and Angelo Bolchi
Department of Biochemistry and Molecular Biology, Italy.

ABSTRACT

The repair of damaged DNA can occur through different pathways, all pointing to the efficient recovery of essential DNA functions. The variety of possible DNA lesions, associated with the redundancy of DNA repair activities, makes DNA repair studies extremely intricate and the specific role of individual proteins very difficult to decipher. In addition to that, damaged DNA structures are structurally very different and exhibit unpredictable localizations within the genome. Conserved proteins or protein domains, which are capable of specific recognition for different non-canonical DNA structures, offer a possible approach to study the mechanisms by which the integrity of genomic functions is monitored within living cells. PARP-type fingers and SAP domains appear to be the first candidates for this analysis because i) they belong to abundant and essential DNA binding proteins; ii) they are very much conserved among eukaryotes; iii) they specifically recognize non-canonical DNA structures; iv) they appear to be involved in chromatin organization; v) they are repetitively found in DNA repair enzymes; iv) they might have partially overlapping functions, despite very different structural features.

Here, we wish to examine the functional significance of PARP-type fingers and SAP domains in the nuclear processes of eukaryotic cells. It would appear as PARP-type fingers and SAP domains are part of DNA binding proteins that, at the same time, cooperate in maintaining correct conformations of specific chromatin domains and signal the presence of damaged DNA structures at the same time.

* Correspondence concerning this article should be addressed to Stefania Petrucco Tel: +39 0521 905149; Fax: +39 0521 905151; e-mail: petrucco@unipr.it.

Introduction

In the recent past it has become apparent that proteins can distinguish specific DNA primary structures, namely specific nucleotide sequences, that are embedded in a roughly invariable DNA secondary structure, namely the B-form DNA. A great number of protein domains capable of sequence-specific DNA recognition have been characterized and, indeed, most transcription factors have been isolated exploiting their sequence-specific DNA binding activities (Ho et al., 2006) and references therein).

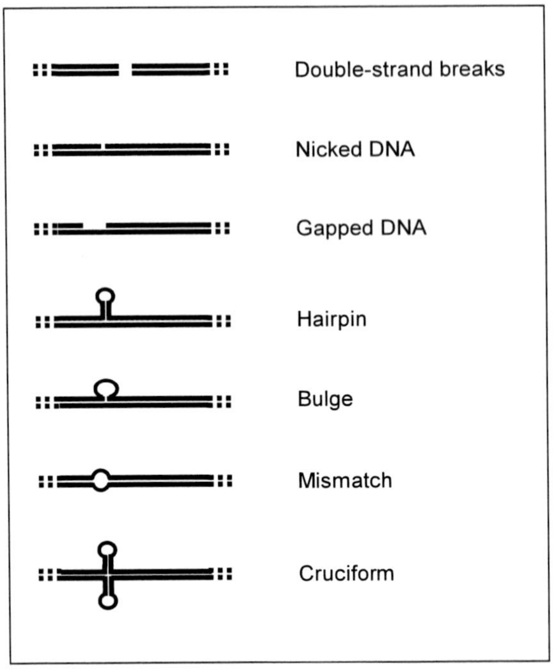

Figure 1. Schematic representaion of frequently observed non-canonical DNA structures.

In most cases, sequence-specific DNA binding is B-form specific and it can be abolished when the recognized sequence is included in non-canonical DNA structures (Figure 1). As an example, restriction enzymes efficiently cut B form-DNA at the corresponding restriction sites, but they are inactive when the same sites are present in single stranded DNAs or include nicks or are folded in hairpin loops. Similarly, transcription factors recognize their cognate sequence in a B- form, double helical context. Therefore, sequence-specific recognition, which has a pivotal role in genomic functions, basically requires the DNA to be in a B-form helical structure.

However, are non-B DNA structures expected to be present in the living cells and, in such cases, do they have specific roles in DNA functions?

Nicks, gaps, double strand breaks, cruciforms, curvatures, bulges of the double strand DNA, all of which can be physiologically produced during DNA transactions or following DNA damages, clearly represent non-canonical DNA secondary structures that can accumulate within the nuclei (Bacolla and Wells, 2004; Chou et al., 2003; Leach, 1994). In addition, high or low levels of DNA supercoiling, e.g. generated by the replication fork or during RNA transcription, can promote structural deformation of the DNA double helix (Blot et al., 2006; Champoux, 2001; Corbett and Berger, 2004; Leonard and Patient, 1991; Ljungman and Hanawalt, 1992). Recent findings, indicating that RNA transcription covers much wider fractions of genomic DNA than previously expected, still broaden the importance of topological control and, hence of non-B DNA structures in living cells (Bertone et al., 2004; Carninci et al., 2005; Cawley et al., 2004).

It is thus clear that DNA secondary structures continuously form within the nuclei and that they must be controlled in order to preserve DNA functions.

Two major systems are known to be involved in the control of undesirable DNA secondary structures, the first concerning the topology of the DNA, the second dedicated to the repair of damaged DNA.

DNA topology is under the control of essential enzymes, namely DNA topoisomerases, which prevent the build-up of superhelical strain during DNA transactions. Eukaryotic DNA topoisomerases transiently cleave and religate DNA strands, via a 3' phopshotyrosyl enzyme-DNA intermediate, which leads to the production of 3'phosphate by-products in the case of abortive reactions. DNA topoisomerases, therefore, are essential enzymes in all organisms, but can themselves be sources of DNA lesions (Champoux, 2001; Corbett and Berger, 2004).

The repair of damaged DNA relies on complex and articulate pathways that can be grouped in three major types: the first one includes repair responses to the different single-strand DNA damages (SSD), the second one is dedicated to the resolution of double strand DNA breaks (DSBs) and the third one is responsible for the removal of mismatches (MM) that are generated during inaccurate DNA replication (de Laat et al., 1999; Essers et al., 2006; Harfe and Jinks-Robertson, 2000; Hoeijmakers, 2001; Wyman and Kanaar, 2006).

To accomplish this task, DNA repair pathways make use of evolutionary conserved enzymes, such as repair-specific DNA polymerases and ligases, DNA kinases and phosphatases, eso- and endonucleases, helicases, DNA glycosylases, as well as the enzymes for homologous and non-homologous recombination processes.

The efficient DNA repair *in vivo,* however, requires additional functions, which integrate and stimulate the activity of the repair enzymes described above. These include DNA topoisomerases and other proteins of known functions, such as single strand DNA sensors, scaffold proteins, DNA-dependent protein kinases. Beside that, specific protein domains are recurrently associated with DNA repair processes in many organisms, but their precise role in the DNA repair machinery is not fully understood ((Hoeijmakers, 2001) and references therein).

PARP-type fingers and SAP domains provide elucidating examples of conserved protein domains that are presumed to have a general role as chromatin organizers and also to function as specific sensors of damaged DNA.

PARP-type Fingers in Poly(ADP-ribosyl)ation Polymerases (PARPs)

PARP-type fingers own their name to the nuclear enzymes where they were identified for the first time, namely, poly(ADP-ribosyl)ation polymerases (PARPs), (Cherney et al., 1987; Suzuki et al., 1987).

PARPs represent a family of multidomain eukaryotic enzymes that catalyze the reversible NAD+dependent poly(ADP-ribosyl)ation of nuclear proteins. Auto and hetero-modification by PARPs is believed to generally reduce affinity for the DNA, due to the negative charges of added ADP-ribosyl groups, and to induce the release of PARP-modified proteins form the DNA ((Ame et al., 2004) and references therein).

PARP-1 is the most abundant and conserved member of the PARP family and it is characterized by the presence of two unusually long zinc-binding motives positioned at the N-terminus of the protein ((D'Amours et al., 1999; de Murcia and Menissier de Murcia, 1994); Figure 2). It has been shown that these two motives are structurally related to the zinc finger fold of glucocorticoid receptors and provide DNA binding functions (Ame et al., 2004; Gradwohl et al., 1987; Gradwohl et al., 1990; Ikejima et al., 1990). Interestingly, the PARP zinc fingers appeared to specifically recognize strand breaks and, based on that, they were classified as DNA nick sensors (de Murcia and Menissier de Murcia, 1994; Menissier-de Murcia et al., 1989). Further studies have subsequently demonstrated that, in addition to nicks, other DNA structures can be recognized, provided that a flexible joint is present on the double stranded DNA (Galande and Kohwi-Shigematsu, 1999; Gradwohl et al., 1987; Oei et al., 1994; Sastry and Kun, 1990; Soldatenkov et al., 2002).

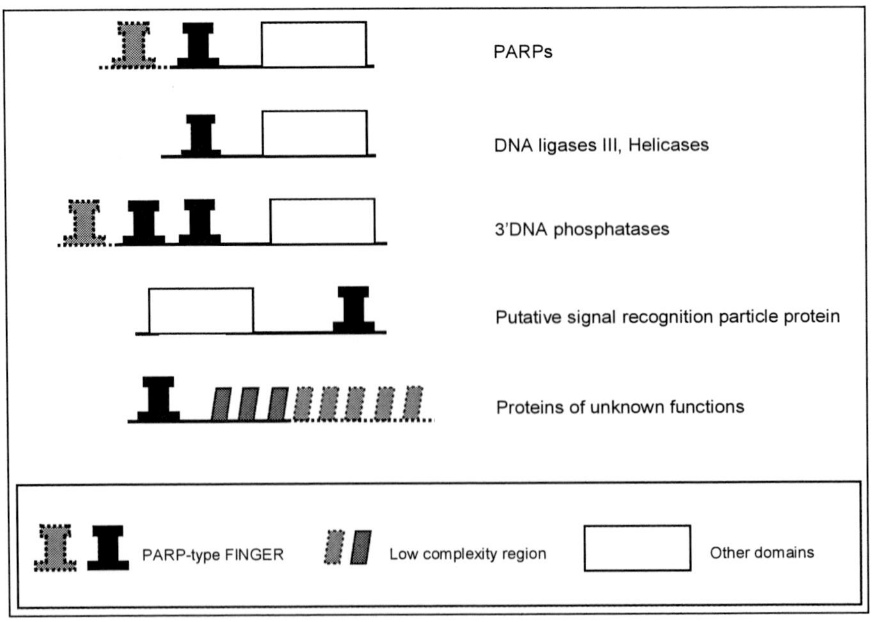

Figure 2. Domain organization of PARP-type fingered proteins. Shading indicates domains that might be absent in the protein type indicated on the right.

PARP-type fingers have a pivotal role in directing protein poly(ADP-ribosyl)ation by PARP-1 since the enzyme becomes active only upon DNA recognition by its fingers. The result is that PARP-1 can modify nuclear proteins, including histones and PARPs themselves, in proximity to DNA sites that are bound by PARP-type fingers (D'Amours et al., 1999; de Murcia and Menissier de Murcia, 1994; Kim et al., 2005).

The amount of protein modification by PARP-1 depends upon the bound DNA structure, the availability of the co-substrate NAD and the cellular levels of ATP (Kim et al., 2004; Lonskaya et al., 2005). These are very relevant features for regulating PARP-1, since they directly linked poly(ADP-ribosyl)ation to the metabolic status of the cells.

But, in the absence of DNA damage, where is PARP-1 localized *in vivo*? Nucleoli, the nuclear matrix, centromeres have all been identified as preferential sites for PARP-1, and indeed it is conceivable that the repetive nature of the DNA at these sites implies specific DNA conformations that can be recognize by PARP-type fingers (Galande and Kohwi-Shigematsu, 1999; Okano et al., 2003; Saxena et al., 2002). In no case however, it has been clarified whether nuclear specific nuclear localization of PARP-1 directly depends on binding of PARP-1 to the DNA or results from interactions with other DNA-bound proteins.

Important insights into the possible role of PARP-1 as chromatin modifier were recently obtained by the identification of a new DNA template for PARP-type fingers. In fact, Kim et al. have shown that PARP-1 fingers display a specific recognition toward nucleosome-bound DNA and, *in vivo*, PARP-1 appears to be associated with non-transcribed euchromatin regions (Kim et al., 2004). Very few details are currently available about the specific nucleosomal organization that is recognized by PARP-type fingers. However, it has been suggested that nucleosome bound PARP-1 is inactive until a triggering NAD/ATP ratio is reached, or until triggering DNA structures are formed. When NAD level raises, ATP level drops, or DNA damage is produced, the bound enzyme becomes activated and can thus be released from the DNA. This results in a specific modification of chromatin accessibility at regions that were previously bound to PARP-1. On this basis, PARP-1 is now considered a new type of chromatin modifier directly modulated by cellular metabolites.

Protein Distributioin of PARP-type Fingers

PARP-type fingers are not unique to PARPs, and they have been described in other eukaryotic enzymes, that is, animal DNA ligases III and plant 3' DNA phosphatases (Mackey et al., 1999; Petrucco, 2003). Furthermore, PARP-type fingers are also encoded in the genomes of lower eukaryotes. In fungi and protozoa, putative PARP-type fingers appear to be associated with proteins of the nucleic acid metabolism, such as DNA helicases and RNA binding proteins, or to be associated with protein regions of low complexity. The organization of PARP-type fingered proteins suggests that the fingers might have a conserved role in the structural recognition of nuclei acid.

Figure 3 reports sequence and structural features that are common among the different members of the PARP-type finger family, and Table I reports a classification of different protein types that include one or more PARP-type fingers.

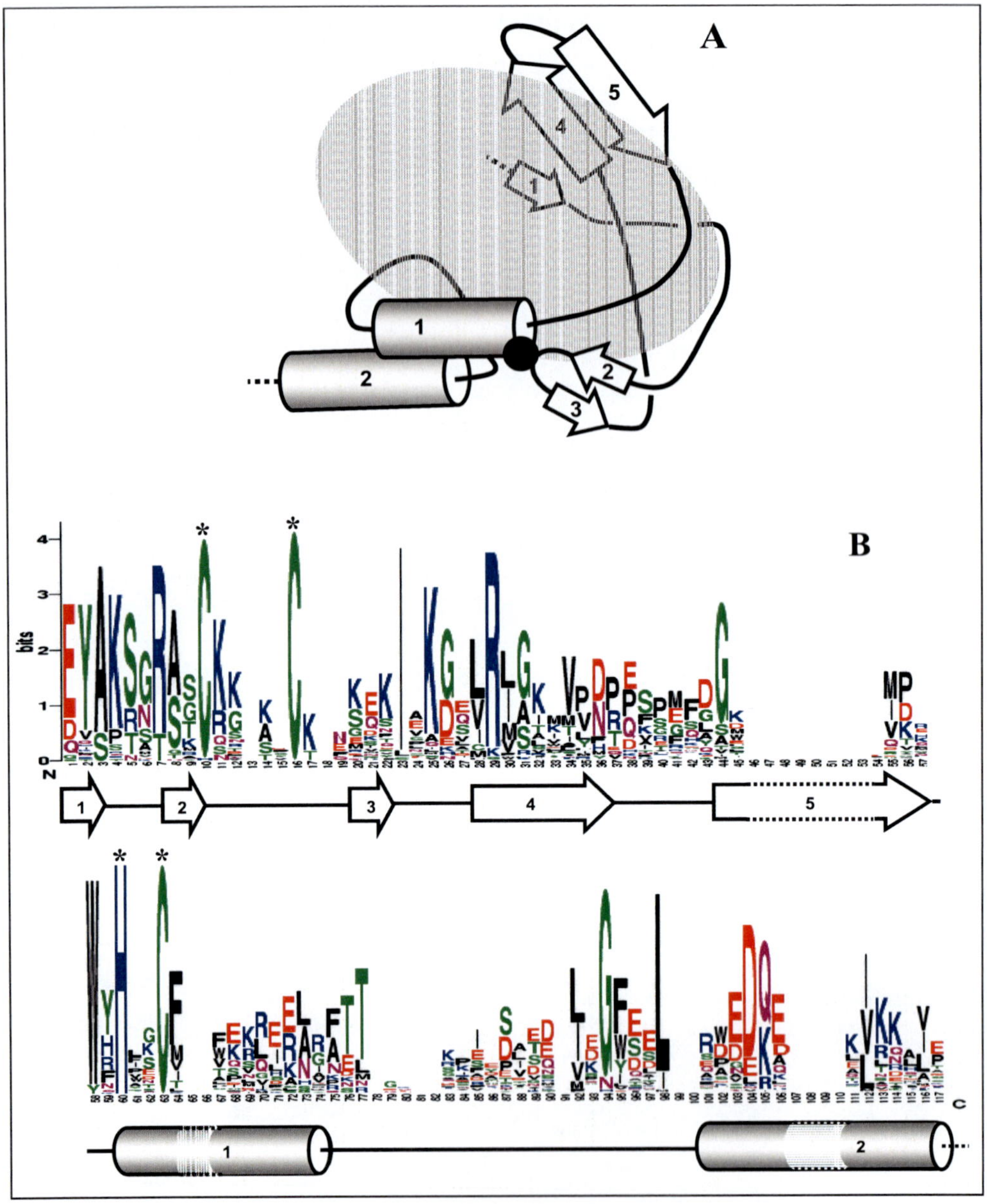

Figure 3. The PARP-type finger fold. A) Conserved structural elements of the PARP-type finger fold are schematically depicted; alpha helixes and beta sheets are numbered, and represented as barrels and arrows, respectively. The black circle indicates the coordinated zinc atom and the shaded area corresponds to the presumed DNA binding region, as mapped by NMR studies (Kulczyk et al., 2004). B) The PARP-type finger sequence alignment was obtained using Clastal W and is graphically represented as a sequence logo (obtained at http://weblogo.berkeley.edu/). Amino acids are coloured according to their chemical properties: polar amino acids (G,S,T,Y,C,Q,N) are green, basic (K,R,H) blue, acidic (D,E) red and hydrophobic (A,V,L,I,P,W,F,M) amino acids are black. Asterisks indicate zinc-coordinating aminoacid residues. Conserved secondary structures are indicated below the sequence logo; shaded areas correspond to secondary structures of variable length within the different PARP-type fingers.

Table I. Functional classification of PARP-type fingered proteins (Pfam database at http://www.sanger.ac.uk/Software/Pfam/)

Protein type	Function	PARP-type finger binding	Nmber of proteins	Organism
PARP	Nuclear protein modification	DNA damage and chromatin	32	Fungi Animals Plants
DNA ligase III	DNA repair	DNA damage	15	Animals
3'DNA phosphatase	DNA repair	DNA damage	4	Plants
DNA helicase	Chromatin unwinding	Not verified	3	Protozoa
Signal recognition particle	RNA binding	Not verified	2	Fungi
Low complexity	Unknown	Not determined	7	Protozoa Fungi

DNA ligases III carry a single N-terminal PARP-type finger, while plant 3' DNA phosphatases bear two or even three such fingers at their N- terminal side (Figure 2). Differently from PARP-1, these two enzymes use DNA as a substrate. In fact DNA ligases III are repair-specific ligases that seal nicks at single strand breaks, and 3' DNA phosphatases are implicated in the specific removal of 3' DNA phosphate, a blocking group that would prevent further mending of DNA strand breaks. It is thus clear that DNA substrate recognition occurs at DNA ligase III and 3' DNA phosphatase active sites and that PARP-type fingers are not required for substrate binding, nor PARP-type fingers appear to regulate the catalytic activities of these two enzymes (Mackey et al., 1999; Petrucco, 2003; Taylor et al., 2000). Hence, PARP-type fingers must supply different functions to these DNA repair proteins. Due to their low abundance, they are not expected to compete with PARP-1 for chromatin binding. On the other hand, should DNA ligase III or 3'DNA phosphatase compete with PARP-1 for damaged DNA sites?

PARP-TYPE FINGERS AS SENSORS OF DNA DAMAGE

It has been proposed that PARP-1, by functioning as a DNA damage sensor, also constitutes a docking site for the assembly of a DNA repair complex, which includes PARP-type fingered DNA ligase III in animal cells and, possibly fingered 3'DNA phosphatase in plant cells (Slupphaug et al., 2003). Indeed, protein interactions between PARP-1 and several proteins of DNA repair complexes are known to occur (e.g. XRCC1, DNA polymerase beta, DNA ligase III, Ku70/80, DNA PKs, mitotic spindle checkpoints, topoisomerases, etc). It must be noted however, that PARP-1, following DNA damage recognition and poly(ADP-ribosyl)ation, has reduced affinity for DNA and, at the same time, increased affinity for several components of the repair compex.

It is difficult to envisage how modified PARP-1, which is released from damaged DNA, could at the same site recrute and stabilize an interacting DNA repair complex. Altough the correct interpretation of this possible contradictory observation would require comprehensive *in vivo* analyses, a series of events that could be compatible with these results can be proposed (Figure 4). As reported above, recent findings directly connect PARP-1 with the control of non-transcribed euchromatin domains (Kim et al., 2004). In case of DNA damage, the binding of PARP-1 to the nucleosome is modified because higher affinity sites are generated at the damaged sites (e.g. strand breaks), and because the nucleosome itself undergoes structural modification due to the DNA lesion. It is possible that, upon damage recognition, enzyme activation and PARP-1 auto-modification, the triggering signal for the repair responses is, in fact, the release of PARP-1 from the nucleosome. PARP-type fingers on DNA repair proteins, such as DNA ligase III or possibly, 3' DNA phosphatase, could then tether the repair complex at damaged DNA sites. At the same time, due to the increased interaction of the repair complex with modified PARP-1, DNA-released PARP-1 remains associated with the damaged euchromatin region.

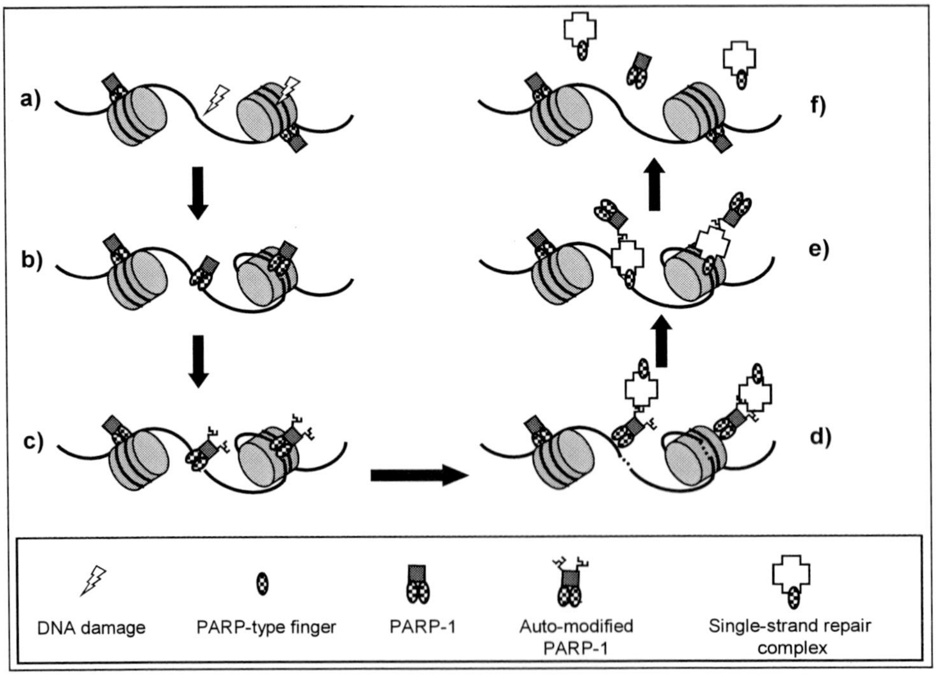

Figure 4. A model for the interplay of PARP-1 and the DNA repair complex on non-transcribed euchromatin. a) PARP-1 is a structural component of non-transcribed euchromatin until DNA damage modifies the DNA structure; b) PARP-1 moves to damaged DNA sites; histone-DNA interactions are modified by the DNA damage and by PARP-1 mobilization; c) Upon induction of its enzymatic activity, PARP-1 becomes auto-poly(ADP-ribosyl)ated; d) Auto-modified PARP-1 is released from the DNA while interacting with the DNA repair complex. e) The DNA repair complex binds to the damaged site and inactivates PARP-1; the enzyme PARG (not depicted) reverses PARP-1 modification. f) The DNA damage is mended, the DNA repair complex is released, PARP-1 regains access to its nucleosomal site. Protein sets interacting with modified and/or non-modified PARP-1 are not shown, neither are histone modifications.

Signalling DNA damages via the release of chromatin bound PARP-1 has important implications in genome surveillance. i) While transcribed euchromatin is directly scanned for DNA damaged by the transcription apparatus itself, non-transcribed euchromatin appears to use PARP-1 DNA binding properties for a continuous control of its integrity. ii) PARP-type fingers on DNA repair proteins, such as DNA ligase III or, possibly 3' DNA phosphatase, do not compete for damaged site with PARP-1, since auto-modified PARP-1 has low affinity for the DNA. iii) The direct interaction of auto-modified PARP-1 with the repair complex has a dual role, namely it inactivates PARP-1 enzymatic activity (Masson et al., 1998) and it maintains PARP-1 in the proximity of the damaged DNA site. As a consequence, PARP-1 is efficiently reverted into its DNA-binding conformation by a poly(ADP-ribose) glycohydrolase (PARG) (Davidovic et al., 2001) and PARP-1 bound euchromatin is rapidly restored, following DNA repair.

The presence of PARP-type fingers in two repair enzymes of single-strand DNA damage, and their absence in proteins of other DNA repair complexes, suggests a specific role of PARP-type fingers in this kind of DNA lesions. Phosphate specific 3'DNA phosphesterases are exclusively found in eukaryotic organisms, as if eukaryotes, but not prokaryotes, are prone to accumulate DNA 3'blocking phosphates (Deshpande and Wilson, 2004). A possible explanation resides in the mechanism of action of eukaryotic topoisomerases, which form 3'phopshotyrosyl enzyme-DNA intermediates, and produce 3'phosphate by-product following abortive reactions. This interpretation would directly link 3'DNA phosphatases, plant ZDP included, with eukaryotic topoisomerases functions. More in general, the amount of DNA secondary structures, as hairpins or cruciforms, depends on the torsional stress imposed on the DNA. Abortive cleavage by topoisomerases, beyond leaving a nicked DNA strand, may improperly release the local DNA torsion. Possibly, PARP-type fingers either on the 3'DNA phosphatases of putative plant SSR complexes, or on the DNA ligases III of animal SSR complexes, limit the unwounding of the DNA by stabilizing local non-B DNA structures in the proximity of broken DNA ends. This, in turn, could be important to control the degradation processes form the broken DNA strands and might also control the amount of flapping DNA which is prone to initiate unwanted recombination events. Noteworthy, broken DNA ends remain accessible to repair since both, DNA ligases III and 3'DNA phosphatases, appear to be perfectly capable at modifing free DNA ends (ligating nicks and removing 3'DNA phosphates, respectively) where PARP-type fingers are bound. Specific helicases, such as Werner protein, which are also known to be part of the SSR repair complex, could eventually partecipate in the final resolution of unwanted non-B DNA structures.

SAP Domains

Given the recurrent finding of PARP-1 in *in vivo* DNA transactions, it came out as a surprise that PARP-1 knockout mice were deficient in DNA repair processes by vital (Shall and de Murcia, 2000). Such result indicates that either PARP-1 is not an essential enzyme or that its functions are redundant. In search for additional PARP activities within the cells that could vicariate PARP-1 functions, it become clear that a family of PARP proteins exists in higher eukaryots, all sharing a highly conserved poly(ADP-ribosylation) domain (Ame et al.,

2004). One member of this family, termed PARP-2 appears to be the functional homolog of PARP-1. This is because PARP-2 is the only other PARP enzyme whose activity is stimulated by damaged DNA, and because PARP-2 can complement PARP-1 functions in knockout mices. In fact, while PARP-1 or PARP-2 deleted mice are viable, double mutant mices lacking both, PARP-1 and PARP-2 are embryonic lethal (Menissier de Murcia et al., 2003). It was thus deduced that PARP-1 and PARP-2 provide redundant essential functions and that either PARP-1 or PARP-2 must be expressed in mammalian cells.

The major difference between PARP-1 and PARP-2 reside in their N-terminal DNA binding region since PARPs-2 possess two SAP domains in place of the two PARP-type fingers of PARP-1, which similarly provide DNA binding and enzyme activation functions (Ame et al., 1999).

The SAP domain is a 35 aminoacid-long motif that is highly conserved in a wide variety of chromatin-associated proteins. Looking at the structural features of the SAP domain it is immediatly evident that there are no sequence nor structural similarities with PARP-type fingers (Figure 5). On the other hand, and analogously to PARP-type fingers, SAP domains are present in proteins implicated in DNA metabolisms as well as in polypeptides bearing long stretches of low complexity regions (Table II). It is worthwhile noting that this last group of SAP-bearing proteins include several transcritpion factors (Figure 6), suggesting that similarly organized PARP-type fingered proteins might also function as transcription factors (see Figure 2).

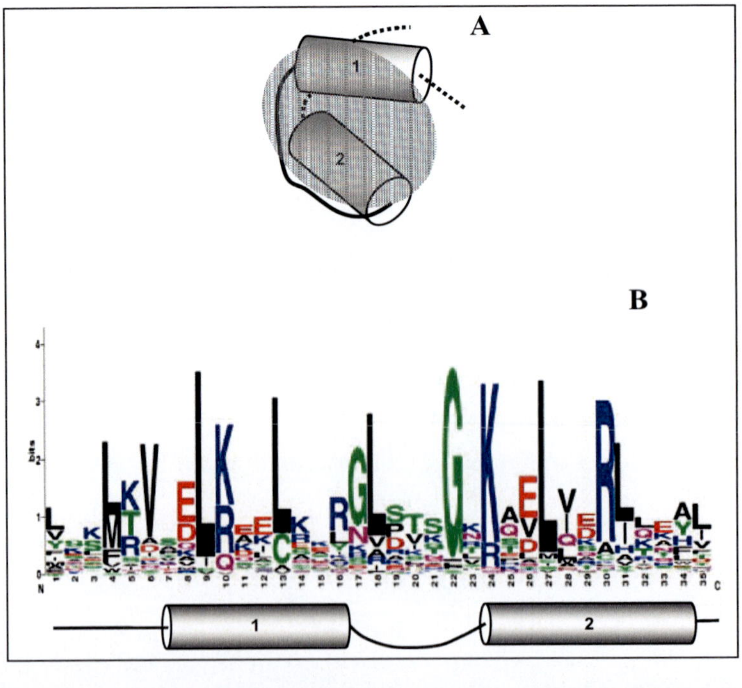

Figure 5. The SAP domain. A) The SAP domain structure is schematized as described in Figure 3. The shaded area indicates the proposed DNA binding surface (Zhang et al., 2001). B) The sequence logo with conserved secondary structures of the SAP domain is as described in Figure 3.

Table II. Functional classification of proteins bearing SAP domains (Pfam database at http://www.sanger.ac.uk/Software/Pfam/)

Protein type	Function	SAP domain binding	Number of proteins	Organism
hnRNP	RNA binding	SARs/MAR	70	Fungi Animals
PIAS	Sumoilation in transcriptional control	SARs/MAR	59	Protozoa Fungi Animals Plants
RAD18	Ubiquitination in DNA repair	Not verified	27	Fungi Animals
Ku70	DNA end binding	Not verified	26	Fungi Animals Plants
Transcription factors	Transcription regulation	Not verified	19	Fungi Animal Plants
Exo/endonuclease	DNA repair	DNA damage	18	Protozoa Fungi Animals Plants
Splicing factors	RNA splicing	Not verified	10	Fungi Animals
PARP	Nuclear protein modification	DNA damage	3	Animals Plants
Peptidase	Ubiquitin pathway	Not determined	1	Virus Fungi
3' DNA phosphatase	DNA repair	Not verified	1	Fungi
Methyltransferase	Ubiquinone and menaquinone biosynthesis	Not determined	1	Fungi
Nucleotide-binding protein	Putative energy status sensor	Not determined	1	Animal
Low complexity	Unknown	Not determined	120	Protozoa Fungi Animals Plants

The SAP domain can specifically bind A-T-rich chromosomal regions named SAR/MARs (matrix/scaffold attachment regions), that represent chromatin organizing centres, where other abundant components of the chromatin, such as toposiomerase II, laminin II, HMGs etc, are also bound (Gohring et al., 1997). It has been hypothesized that A-

T rich regions can support a specific DNA conformation that dictates their functional role as chromatin organizers *in vivo*. SAP domains are probably involved in the recognition of this DNA conformation (Jenuwein et al., 1997).

Interestingly, A-T rich DNA is also recognized by PARP-type fingers *in vitro* (Galande and Kohwi-Shigematsu, 1999), while the *in vivo* binding of PARP-type fingers at these chromatin sites has not been tested.

Hence, on the basis of its SAR/MARs binding and of its presence in different classes of proteins involved in the interaction with unusual structures of the nucleic acids, it can be concluded that SAP domains represent another domain type with the ability of recognizing DNA in a structure-specific manner.

Based on the functional redundancy between PARP-1 and PARP-2, it can be anticipated that PARP-type fingers and SAP domains can recognize similar DNA structures *in vivo*. The DNA binding properties of PARP-type fingers have been extensively tested *in vitro*, leading to the hypothesis that they can generally recognize sites of enhanced flexibility of the double helix, including nicks, gaps, cruciforms, bended DNA and hairpins (Gradwohl et al., 1987; Lonskaya et al., 2005; Sastry and Kun, 1990; Soldatenkov et al., 2002). On the other hand, very limited information is available about the DNA binding specificity of SAP domains.

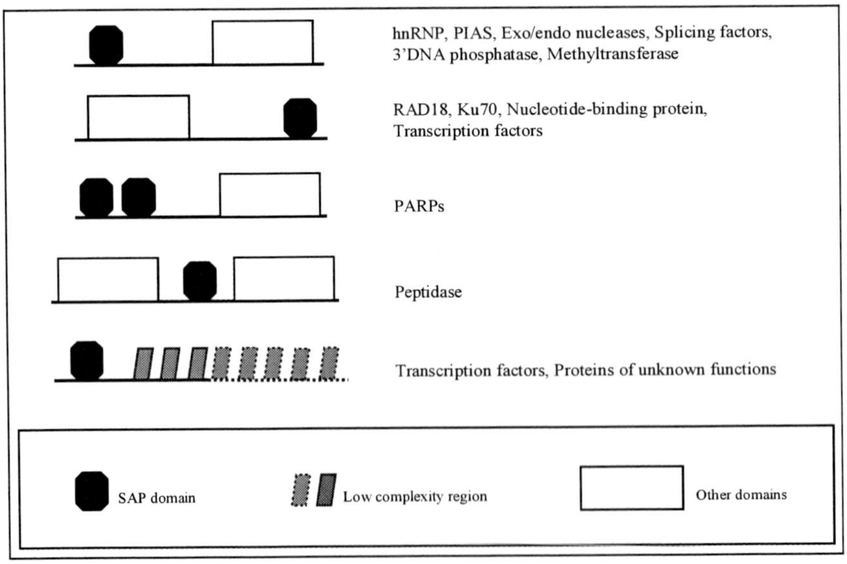

Figure 6. Domain organization of proteins bearing SAP domains. Shading indicates domains that might be absent in the protein types indicated on the right.

Certainly, a range of DNA binding modules (e.g. the AT hook domain, the Bright domain, etc) are known to both, recognize non-canonical DNA structures similar to the ones bound by PARP-type fingers, and play a role in the organization of chromatin structure (Aravind and Landsman, 1998; Baxevanis and Landsman, 1995; Gregory et al., 1996). However, a major difference among these domains and PARP-type fingers or SAP domains, is that only the lattest two repetitively occur as the DNA binding domains of proteins involved in the DNA repair. This evidence underlines the specific role of PARP-type fingers and SAP domains in the chromatin organization and genome surveillance. In this context, it is interesting to note that PARP enzymes might not be the sole example of DNA repair enzymes where PARP-type fingers and SAP domains have interchangeable functions. In fact, the Dictostlylium genome encodes a 3'DNA phosphatases, which is predicted to contain SAP domain, in place of the PARP-type fingers of the plant 3' DNA phosphatases (accession number Q54U78).

Whichever the real meaning of the functional equivalence between PARP-type fingers and SAP domains, we believe that it is a clear indication that the structural recognition of the DNA is more refined than previously thought and that subsets of DNA structures can have very specific functional roles within the eukaryotic cells.

Conclusion

PARP-type fingers and SAP domains constitute two types of protein modules, bearing specific recognition properties for non-canonical nucleic acid structures. DNA binding by PARP-type fingers or SAP domains does not appear to be essential for *in vitro* DNA transactions suggesting that structural recognition of the DNA is mostly important on chromatinized templates. Functional *in vivo* evidence indicates that these two structure-specific DNA binding modules have redundant and essential roles in higher eukaryotes. PARP-type fingers, as well as SAP domains, appear to be associated with different as well as identical protein types implicated in nucleic acid metabolism. Also, PARP-type fingers and SAP domains are frequently identified in polypeptides bearing extended low complexity regions, which could represent sites of protein-protein interactions and suggest that PARP-type fingers or SAP domain could serve for bridging interacting proteins at specific DNA structures.

The specific role of PARP-1 and PARP-2 in the DNA repair processes and as chromatin modifers indicate that structural recognition of DNA could provide a powerful tool for organizing functional chromatin and for sensing DNA defects.

Future work is expected to clarify these points and to gain insights into the DNA anchors available to eukaryotic cells for organizing the intricate structures of chromatin.

References

Ame, J. C., Rolli, V., Schreiber, V., Niedergang, C., Apiou, F., Decker, P., Muller, S., Hoger, T., Menissier-de Murcia, J., and de Murcia, G. (1999). PARP-2, A novel mammalian DNA damage-dependent poly(ADP-ribose) polymerase. *J Biol Chem 274*, 17860-17868.

Ame, J. C., Spenlehauer, C., and de Murcia, G. (2004). The PARP superfamily. Bioessays *26*, 882-893.

Aravind, L., and Landsman, D. (1998). AT-hook motifs identified in a wide variety of DNA-binding proteins. *Nucleic Acids Res 26*, 4413-4421.

Bacolla, A., and Wells, R. D. (2004). Non-B DNA conformations, genomic rearrangements, and human disease. *J Biol Chem 279*, 47411-47414.

Baxevanis, A. D., and Landsman, D. (1995). The HMG-1 box protein family: classification and functional relationships. *Nucleic Acids Res 23*, 1604-1613.

Bertone, P., Stolc, V., Royce, T. E., Rozowsky, J. S., Urban, A. E., Zhu, X., Rinn, J. L., Tongprasit, W., Samanta, M., Weissman, S., *et al.* (2004). Global identification of human transcribed sequences with genome tiling arrays. *Science 306*, 2242-2246.

Blot, N., Mavathur, R., Geertz, M., Travers, A., and Muskhelishvili, G. (2006). Homeostatic regulation of supercoiling sensitivity coordinates transcription of the bacterial genome. *EMBO Rep 7*, 710-715.

Carninci, P., Kasukawa, T., Katayama, S., Gough, J., Frith, M. C., Maeda, N., Oyama, R., Ravasi, T., Lenhard, B., Wells, C., *et al.* (2005). The transcriptional landscape of the mammalian genome. *Science 309*, 1559-1563.

Cawley, S., Bekiranov, S., Ng, H. H., Kapranov, P., Sekinger, E. A., Kampa, D., Piccolboni, A., Sementchenko, V., Cheng, J., Williams, A. J., *et al.* (2004). Unbiased mapping of transcription factor binding sites along human chromosomes 21 and 22 points to widespread regulation of noncoding RNAs. *Cell 116*, 499-509.

Champoux, J. J. (2001). DNA topoisomerases: structure, function, and mechanism. *Annu Rev Biochem 70*, 369-413.

Cherney, B. W., McBride, O. W., Chen, D. F., Alkhatib, H., Bhatia, K., Hensley, P., and Smulson, M. E. (1987). cDNA sequence, protein structure, and chromosomal location of the human gene for poly(ADP-ribose) polymerase. *Proc Natl Acad Sci U S A 84*, 8370-8374.

Chou, S. H., Chin, K. H., and Wang, A. H. (2003). Unusual DNA duplex and hairpin motifs. *Nucleic Acids Res 31*, 2461-2474.

Corbett, K. D., and Berger, J. M. (2004). Structure, molecular mechanisms, and evolutionary relationships in DNA topoisomerases. *Annu Rev Biophys Biomol Struct 33*, 95-118.

D'Amours, D., Desnoyers, S., D'Silva, I., and Poirier, G. G. (1999). Poly(ADP-ribosyl)ation reactions in the regulation of nuclear functions. *Biochem J 342*, 249-268.

Davidovic, L., Vodenicharov, M., Affar, E. B., and Poirier, G. G. (2001). Importance of poly(ADP-ribose) glycohydrolase in the control of poly(ADP-ribose) metabolism. *Exp Cell Res 268*, 7-13.

de Laat, W. L., Jaspers, N. G., and Hoeijmakers, J. H. (1999). Molecular mechanism of nucleotide excision repair. *Genes Dev 13*, 768-785.

de Murcia, G., and Menissier de Murcia, J. (1994). Poly(ADP-ribose) polymerase: a molecular nick-sensor. *Trends Biochem Sci 19*, 172-176.

Deshpande, R. A., and Wilson, T. E. (2004). Identification of DNA 3'-phosphatase active site residues and their differential role in DNA binding, Mg2+ coordination, and catalysis. *Biochemistry 43*, 8579-8589.

Essers, J., Vermeulen, W., and Houtsmuller, A. B. (2006). DNA damage repair: anytime, anywhere? *Curr Opin Cell Biol 18*, 240-246.

Galande, S., and Kohwi-Shigematsu, T. (1999). Poly(ADP-ribose) polymerase and Ku autoantigen form a complex and synergistically bind to matrix attachment sequences. *J Biol Chem 274*, 20521-20528.

Gohring, F., Schwab, B. L., Nicotera, P., Leist, M., and Fackelmayer, F. O. (1997). The novel SAR-binding domain of scaffold attachment factor A (SAF-A) is a target in apoptotic nuclear breakdown. *Embo J 16*, 7361-7371.

Gradwohl, G., Mazen, A., and de Murcia, G. (1987). Poly(ADP-ribose) polymerase forms loops with DNA. *Biochem Biophys Res Commun 148*, 913-919.

Gradwohl, G., Menissier de Murcia, J. M., Molinete, M., Simonin, F., Koken, M., Hoeijmakers, J. H., and de Murcia, G. (1990). The second zinc-finger domain of poly(ADP-ribose) polymerase determines specificity for single-stranded breaks in DNA. *Proc Natl Acad Sci U S A 87*, 2990-2994.

Gregory, S. L., Kortschak, R. D., Kalionis, B., and Saint, R. (1996). Characterization of the dead ringer gene identifies a novel, highly conserved family of sequence-specific DNA-binding proteins. *Mol Cell Biol 16*, 792-799.

Harfe, B. D., and Jinks-Robertson, S. (2000). DNA mismatch repair and genetic instability. *Annu Rev Genet 34*, 359-399.

Ho, S. W., Jona, G., Chen, C. T., Johnston, M., and Snyder, M. (2006). Linking DNA-binding proteins to their recognition sequences by using protein microarrays. *Proc Natl Acad Sci U S A 103*, 9940-9945.

Hoeijmakers, J. H. (2001). Genome maintenance mechanisms for preventing cancer. *Nature 411*, 366-374.

Ikejima, M., Noguchi, S., Yamashita, R., Ogura, T., Sugimura, T., Gill, D. M., and Miwa, M. (1990). The zinc fingers of human poly(ADP-ribose) polymerase are differentially required for the recognition of DNA breaks and nicks and the consequent enzyme activation. Other structures recognize intact DNA. *J Biol Chem 265*, 21907-21913.

Jenuwein, T., Forrester, W. C., Fernandez-Herrero, L. A., Laible, G., Dull, M., and Grosschedl, R. (1997). Extension of chromatin accessibility by nuclear matrix attachment regions. *Nature 385*, 269-272.

Kim, M. Y., Mauro, S., Gevry, N., Lis, J. T., and Kraus, W. L. (2004). NAD+-dependent modulation of chromatin structure and transcription by nucleosome binding properties of PARP-1. *Cell 119*, 803-814.

Kim, M. Y., Zhang, T., and Kraus, W. L. (2005). Poly(ADP-ribosyl)ation by PARP-1: 'PAR-laying' NAD+ into a nuclear signal. *Genes Dev 19*, 1951 1967.

Kulczyk, A. W., Yang, J. C., and Neuhaus, D. (2004). Solution structure and DNA binding of the zinc-finger domain from DNA ligase IIIalpha. *J Mol Biol 341*, 723-738.

Leach, D. R. (1994). Long DNA palindromes, cruciform structures, genetic instability and secondary structure repair. *Bioessays 16*, 893-900.

Leonard, M. W., and Patient, R. K. (1991). Evidence for torsional stress in transcriptionally activated chromatin. *Mol Cell Biol 11*, 6128-6138.

Ljungman, M., and Hanawalt, P. C. (1992). Localized torsional tension in the DNA of human cells. *Proc Natl Acad Sci U S A 89*, 6055-6059.

Lonskaya, I., Potaman, V. N., Shlyakhtenko, L. S., Oussatcheva, E. A., Lyubchenko, Y. L., and Soldatenkov, V. A. (2005). Regulation of poly(ADP-ribose) polymerase-1 by DNA structure-specific binding. *J Biol Chem 280*, 17076-17083.

Mackey, Z. B., Niedergang, C., Murcia, J. M., Leppard, J., Au, K., Chen, J., de Murcia, G., and Tomkinson, A. E. (1999). DNA ligase III is recruited to DNA strand breaks by a zinc finger motif homologous to that of poly(ADP-ribose) polymerase. Identification of two functionally distinct DNA binding regions within DNA ligase III. *J Biol Chem 274*, 21679-21687.

Masson, M., Niedergang, C., Schreiber, V., Muller, S., Menissier-de Murcia, J., and de Murcia, G. (1998). XRCC1 is specifically associated with poly(ADP-ribose) polymerase and negatively regulates its activity following DNA damage. *Mol Cell Biol 18*, 3563-3571.

Menissier de Murcia, J., Ricoul, M., Tartier, L., Niedergang, C., Huber, A., Dantzer, F., Schreiber, V., Ame, J. C., Dierich, A., LeMeur, M., *et al.* (2003). Functional interaction between PARP-1 and PARP-2 in chromosome stability and embryonic development in mouse. *Embo J 22*, 2255-2263.

Menissier-de Murcia, J., Molinete, M., Gradwohl, G., Simonin, F., and de Murcia, G. (1989). Zinc-binding domain of poly(ADP-ribose)polymerase participates in the recognition of single strand breaks on DNA. *J Mol Biol 210*, 229-233.

Oei, S. L., Herzog, H., Hirsch-Kauffmann, M., Schneider, R., Auer, B., and Schweiger, M. (1994). Transcriptional regulation and autoregulation of the human gene for ADP-ribosyltransferase. *Mol Cell Biochem 138*, 99-104.

Okano, S., Lan, L., Caldecott, K. W., Mori, T., and Yasui, A. (2003). Spatial and temporal cellular responses to single-strand breaks in human cells. *Mol Cell Biol 23*, 3974-3981.

Petrucco, S. (2003). Sensing DNA damage by PARP-like fingers. *Nucleic Acids Res 31*, 6689-6699.

Sastry, S. S., and Kun, E. (1990). The interaction of adenosine diphosphoribosyl transferase (ADPRT) with a cruciform DNA. *Biochem Biophys Res Commun 167*, 842-847.

Saxena, A., Saffery, R., Wong, L. H., Kalitsis, P., and Choo, K. H. (2002). Centromere proteins Cenpa, Cenpb, and Bub3 interact with poly(ADP-ribose) polymerase-1 protein and are poly(ADP-ribosyl)ated. *J Biol Chem 277*, 26921-26926.

Shall, S., and de Murcia, G. (2000). Poly(ADP-ribose) polymerase-1: what have we learned from the deficient mouse model? *Mutat Res 460*, 1-15.

Slupphaug, G., Kavli, B., and Krokan, H. E. (2003). The interacting pathways for prevention and repair of oxidative DNA damage. *Mutat Res 531*, 231-251.

Soldatenkov, V. A., Chasovskikh, S., Potaman, V. N., Trofimova, I., Smulson, M. E., and Dritschilo, A. (2002). Transcriptional repression by binding of poly(ADP-ribose) polymerase to promoter sequences. *J Biol Chem 277*, 665-670.

Suzuki, H., Uchida, K., Shima, H., Sato, T., Okamoto, T., Kimura, T., and Miwa, M. (1987). Molecular cloning of cDNA for human poly(ADP-ribose) polymerase and expression of its gene during HL-60 cell differentiation. *Biochem Biophys Res Commun 146*, 403-409.

Taylor, R. M., Whitehouse, C. J., and Caldecott, K. W. (2000). The DNA ligase III zinc finger stimulates binding to DNA secondary structure and promotes end joining. *Nucleic Acids Res 28*, 3558-3563.

Wyman, C., and Kanaar, R. (2006). DNA Double-Strand Break Repair: All's Well That Ends Well. *Annu Rev Genet*.

Zhang, Z., Zhu, L., Lin, D., Chen, F., Chen, D. J., and Chen, Y. (2001). The three-dimensional structure of the C-terminal DNA-binding domain of human Ku70. *J Biol Chem 276*, 38231-38236.

In: DNA Research Trends
Editor: Marius P. Young, pp. 177-190

ISBN: 978-1-60021-646-6

Chapter VIII

EFFECT OF ANTISENSE VEGF cDNA TRANSFECTION ON THE GROWTH OF CHRONIC MYELOID LEUKEMIA K562 CELLS *IN VITRO* AND IN NUDE MICE

Ruan Guo-Rui, Liu Yan-Rong*, Chen Shan-Shan*, Fu Jia-Yu, Chang Yan, Qin Ya-Zhen, Li Jin-Lan, Yu Hong, Wang Hui

Institute of Hematology and People's Hospital, Peking University, China.

ABSTRACT

To further elucidate the role of vascular endothelial growth factor (VEGF) in the pathogenesis of chronic myeloid leukemia (CML), we transfected K562 cells with a $VEGF_{121}$cDNA sense vector (S), an antisense (AS) vector or vector (V) alone. The growth of transfected cells was investigated by MTT and colony formation assays, and apoptosis was measured by flow cytometry (FCM) of Annexin-V-FITC/PI dual labeled cells. Transfected cells were subcutaneously transplanted into nude mice and the microvessel density (MVD) of tumor masses was determined by vWF immunohistochemistry staining. We tested the supernatant of different transfected K562 cells against human bone marrow endothelial cells (BMECs), and examined the synergic effects of antisense $VEGF_{121}$cDNA and IFNα or STI571 on the proliferation and apoptosis of K562 cells. We found that K562/AS transfectants exhibited a 49% reduction in VEGF secretion, whereas K562/S transfectants exhibited a 3-fold increase in VEGF secretion, all in comparison to the vector controls. K562 cells transfected with antisense $VEGF_{121}$cDNA showed growth retardation *in vitro*. In transplanted nude mice *in vivo*, transfection of implanted cells with antisense $VEGF_{121}$cDNA resulted in decreased tumor MVD, and increased apoptosis in the presence of IFNα. Taken together, these results

* Correspondence concerning this article should be addressed to Dr. Liu Yan-Rong, Prof. Chen Shan-Shan, Institute of Hematology and People's Hospital, Peking University, 11 Xi-zhi-men South Street, Beijing 100044, China. Tel: +8610-68792787; Fax: +8610-68792787, +8610-68344179; E-mail: celiar@public3.bta.net.cn.

suggest that VEGF may be involved in the pathogenesis of CML through autocrine and paracrine mechanisms, and that anti-VEGF therapy alone or in combination with conventional treatment may be beneficial for CML patients.

Keywords: chronic myeloid leukemia, vascular endothelial growth factor (VEGF), transfection, proliferation, apoptosis.

Abbreviations

VEGF, vascular endothelial growth factor;
CML, chronic myeloid leukemia;
BMEC, bone marrow endothelial cell;
MVD, microvessel density;
EGF, endothelial growth factor;
bFGF, basic fibroblast growth factor;
vWF, von Willebrand factor;
UEA-1, Ulex Europeaus Agglutinin-1.

1. Introduction

Solid tumors that grow beyond 2 mm^3 in size require angiogenesis for maintenance of an adequate blood supply [1]. Tumor angiogenesis is regulated by a number of positive or negative growth factors [2], including the important mediator of angiogenesis, vascular endothelial growth factor (VEGF) [3,4]. In general, the levels of VEGF mRNA and protein expression in human solid tumors are positively correlated with malignant progression [5-8]. Recently, it was shown that VEGF expression is higher in the bone marrow of patients with chronic myeloid leukemia (CML) than in that of normal individuals [9-12]. Our previous study demonstrated that the mean VEGF concentration in plasma from untreated CML patients was higher than that from CML patients post allograft bone marrow transplantation (Allo-BMT) [13]. However, the effect of VEGF on the abnormal proliferation of CML cells has not been elucidated to date. Here, we investigated the effects of autocrine VEGF on CML (K562) cells by using VEGF sense and antisense transfected cells for *in vitro* assays and in nude mice *in vivo*.

2. Materials and Methods

2.1. Cell Culture Conditions

K562 cells were grown in RPMI 1640 (GIBCO) supplemented with 10% fetal calf serum (FCS, GIBCO). BMECs (bone marrow endothelial cells) were kindly provided by Prof. Francisco J Candal (Centers for Disease Control and Prevention, USA) and cultured in

MCDB131 (GIBCO) with 20 ng/ml epidermal growth factor (EGF, GIBCO), 1μg/ml hydrocortisone (Sigma), 10 mmol/L L-Glutamine and 10~15% fetal bovine serum (FBS, GIBCO). Cells were incubated at 37°C in a humid atmosphere with 5% CO_2. Prior to use, BMECs were identified by confirmation of vWF expression (by immunohistochemistry) and UEA-1 binding (by flow cytometry).

2.2. VEGF cDNA Transfection

The full-length cDNAs for the $VEGF_{121}$ sense and antisense vectors were gifts from Prof. Zhou Ai-Ru and Prof. Zhen Jie (Peking University, Beijing). The cDNAs were cloned into the EcoRI/BamHI or HindIII/BamHI sites, respectively, of $pcDNA_3$, a eukaryotic expression vector driven by the human cytomegalovirus promoter. Inserts were confirmed by restriction enzyme analysis and direct sequencing.

$VEGF_{121}$sense (S) and $VEGF_{121}$antisense (AS) vectors, or vector alone (V) were transfected by DOTAP Liposomal Tranfection Reagent according to the manufacturer's protocol (Boehringer Mannheim, Germany). Briefly, 1×10^6 K562 cells were seeded into a 35 mm tissue culture dish and incubated in 0.8 ml serum-free RPMI 1640 medium with 15 μl liposome and 10 μg plasmid DNA for 12 hours. The medium was then replaced with 2 ml fresh 10% FCS RPMI 1640 medium, and geneticin (500~1000μg/ml; G418, GIBCO) was added after 72 hours. The cells were fed every three days with fresh medium. After 3 weeks, G418-resistant colonies were transferred separately to individual wells of a 96-well plate, and colonies were expanded and identified.

2.3. Preparation of cDNA and reverse Transcription-polymerase Chain Reaction (RT-PCR)

The various transfected K562 cells (above) were washed twice in phosphate-buffered saline (PBS), centrifuged and collected. Total cellular RNA was extracted using the TRIzol reagent (Invitrogen, USA) as described by the manufacturer. One microgram of RNA was used for cDNA synthesis employing avian myeloblastosis virus (AMV) reverse transcriptase and oligo dT as the primer. cDNA was amplified with specific primers for VEGF, and β_2-microglobulin was used as an internal control to confirm successful cDNA synthesis. For VEGF and β_2-microglobulin, one round of amplification was performed under the following conditions: 35 cycles, 1.5 minutes at 94°C, 3 minutes at 60°C, and 4 minutes at 72°C. Primer sequences were as follows: VEGF (516 bp) sense, 5'-TCG GGC CTC CGA AAC CAT GA-3'; antisense, 5'-CCT GGT GAG AGA TCT GGT TC-3'; β_2-microglobulin (β_2-MG; 120bp) sense, 5'-ACC CCC ACT GAA AAA GAT GA-3'; antisense, 5'-ATC TTC AAA CCT CCA TGA TG-3' [13,17]. PCR products were separated on 8% polyacrylamide gels visualized by silver staining, the band intensities on gel photographs were directly quantitated by computer-assisted image analysis (Four Stars Digital Image System, Shanghai), and the target/control ratios were calculated.

2.4. Enzyme-linked Immunosorbent Assay (ELISA) of VEGF and bFGF

The VEGF ELISA kit was purchased from R & D Systems (USA). The protein levels of VEGF in the cell culture supernatants from different transfected K562 cells were assayed by ELISA kit, following the manufacturer's instructions. Transfected K562 cells in the phase of logarithmic growth were washed three times with serum-free medium, and 3×10^4 cells/ml were incubated for 72 hours in RPMI 1640 containing 5% FCS. The supernatants were collected in tubes and frozen at –20°C until use. The protein levels of bFGF in supernatants were also analyzed by ELISA kit (Biosource Company, USA), following the manufacturer's instructions.

2.5. *In Vitro* Growth Rates

Transfected cells (1×10^4 cells/well) were seeded into 96-well plates. MTT assays were performed after 4 (t=0), 24, 48 and 72 hours incubation. Briefly, MTT solution (20 μl of a 5 mg/ml solution) was added to the wells, which were then incubated for an additional 2 hours at 37°C. The reaction was stopped with 100 μl/well of 0.01 M HCl in 10% SDS. The absorbance was determined on a plate reader (Bio-Rad 550, USA) at 570nm. Each assay was repeated 20 times. Growth curves were generated by Excel computer software (Microsoft, USA).

2.6. Neutralization of VEGF Bioactivity

K562/V (1×10^4 cells/well) were seeded into 96-well plates. Different concentrations (5~160 μg/ml) of anti-human VEGF monoclonal antibody (R & D Systems, USA) were used to neutralize VEGF activity in these cells, and after 72 hours incubation, MTT assays were performed as above. Each assay was repeated 5 times. Growth curves were generated by Excel computer software.

2.7. Colony-Formation Assay

Transfected cells were seeded (500 cells/well) in RPMI 1640 containing 10% FCS, 10% horse serum, 80 ng/ml GM-CSF, and 0.7% methylcellulose. Growth units (≥50 cells) were counted after 12 days. Each assay was repeated 3 times.

2.8. Determination of Apoptosis

The effects of Intron A (IFNα-2b, Schering-Plough, USA) or STI571 (signal transduction inhibitor 571, Novartis, Switzerland) on the apoptosis of transfected K562 cells were studied by flow cytometric Annexin-V-FITC/propidium dual labeling technique.

Transfected cells were cultured to a density of 5×10^4 cells/ml in RPMI 1640 containing 10% FCS with 100 U/ml IFNα or 0.1 μmol/L STI571 for 72 hours. The Annexin-V-Fluos kit was purchased from Boehringer Mannheim Company (Germany) and determinations were performed according to the manufacturer's instructions.

2.9. Effects of Culture Supernatants from Different Transfected K562 Cell Lines on BMEC Proliferation

Transfected cells (2.5×10^5 cells/ml) were seeded into RPMI 1640 medium containing 5% FCS for 24 hours. The supernatants were harvested and frozen at –20°C. BMECs were plated (6×10^3 cells/well) in 96-well plates and incubated with BMEC medium (as described in the section of the cell culture conditions) for 24 hours, after which the culture medium was replaced with 25% culture supernatants from transfected cells. As a positive control, one supernatant (from K562/S cells) was pre-neutralized with 0.1 μg/ml anti-VEGF monoclonal antibody. MTT assays were performed after 72 hours incubation. Each assay was repeated 4 times. Growth curves were generated by Excel computer software.

2.10. *In Vivo* Tumor Analysis

Approximately 10^7 transfected K562 cells resuspended in 100 μl of serum-free cell culture medium were subcutaneously injected into the armpits of female nude mice (4~5 weeks of age). The mice were monitored daily for tumor sizes, and tumor volumes were calculated as length×width×height (mm^3). Tumor-bearing mice were sacrificed on the 14th day and tumor masses were removed, photographed and imbedded in paraffin wax for sectioning. Angiogenesis in solid tumors was evaluated by vWF (Dako, Denmark) immunohistochemistry. MVD was evaluated by light microscopy as previously described [14]. Briefly, 1~3 stained slides per case were evaluated at low magnification (×40 and ×100) to detect the area with the highest MVD (hot spot). Three microscopic fields were then examined in this area at ×250 magnification (each field representing an area of 0.72 mm^2), and the mean MVD value was recorded.

3. Results

RT-PCR results showed that the expression of VEGF mRNA was inhibited in K562 cells transfected with $VEGF_{121}$antisense vector (AS), as compared to K562 cells transfected with vector alone (V), but increased in K562 cells transfected with the $VEGF_{121}$sense vector (S) (Figure 1). Computer-assisted image analysis showed that the ratios of VEGF/β_2-MG band intensities were as follows: K562/V: 0.47±0.05, K562/AS: 0.22±0.10 and K562/S: 1.14±0.92.

The ELISA results showed that in comparison to the vector-only controls, K562/AS cells exhibited a 49% reduction and K562/S cells exhibited a 3-fold increase in VEGF secretion

(K562/AS *vs* K562/V, 183.4± 22.1:370.7± 70.8 pg/ml, P<0.01; K562/S *vs* K562/V, 1616.4±415.6:370.7± 70.8 pg/ml, P<0.01; Figure 2). Up- or down-regulation of bFGF was not observed (data not shown), indicating that antisense or sense VEGF cDNA was specific to VEGF. MTT assays showed that the proliferation of K562/AS cells at 72 hours was reduced as compared to that of K562/V cells (p<0.01), but that the proliferation of K562/S cells at 72 hours was faster than that of K562/V cells (p<0.01, Figure 3). Similar results were obtained using the colony-formation assay (Figure 4). When we verified these results with the positive control anti-VEGF monoclonal antibody, we found that proliferation of K562/V cells was reduced by 5 μg/ml anti-VEGF monoclonal antibody and was completely inhibited by 160 μg/ml anti-VEGF monoclonal antibody (Figure 5). In addition, K562/AS yielded more apoptotic cells than did K562/V and K562/S cultures under the same conditions. Antisense $VEGF_{121}$cDNA transfection in combination with 100 U/ml IFNα increased the apoptosis of K562/AS cells, whereas 0.1 μmol/L STI571 markedly increased the apoptosis of all transfected cells (Table 1).

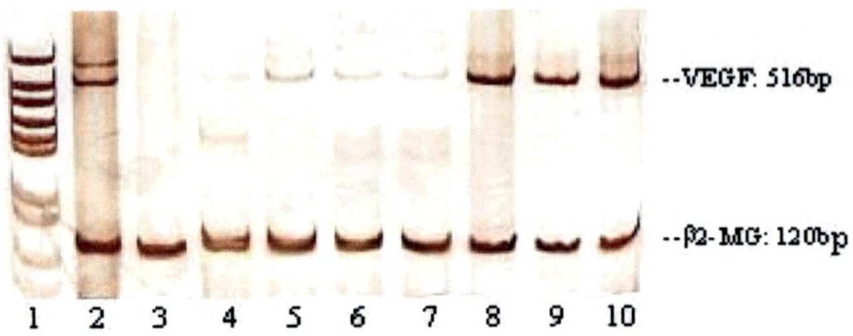

Figure 1. Expression of VEGF mRNA in K562/S and K562/AS cells from different clones, as measured by RT-PCR. Lane 1: Marker. Lane 2: K562/V cells, positive control. Lane 3: Jurkat cells, negative control. Lanes 4~7: K562/AS cells from different clones. Lanes 8~10: K562/S cells from different clones.

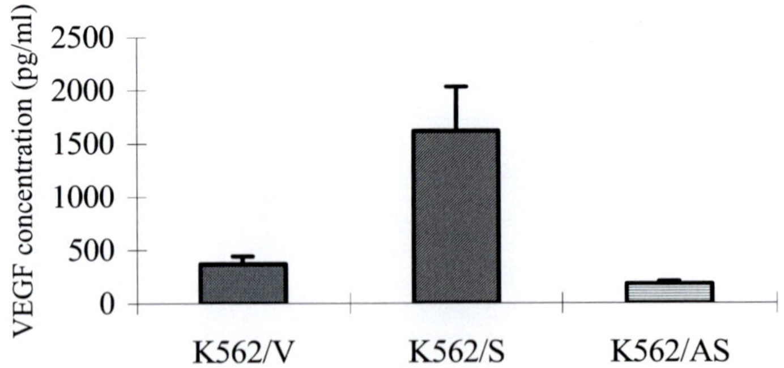

Figure 2. VEGF protein concentrations in the culture supernatants at 72 hours from K562/V, K562/S and K562/AS cells (n=4, P<0.01, single factor analysis of variance).

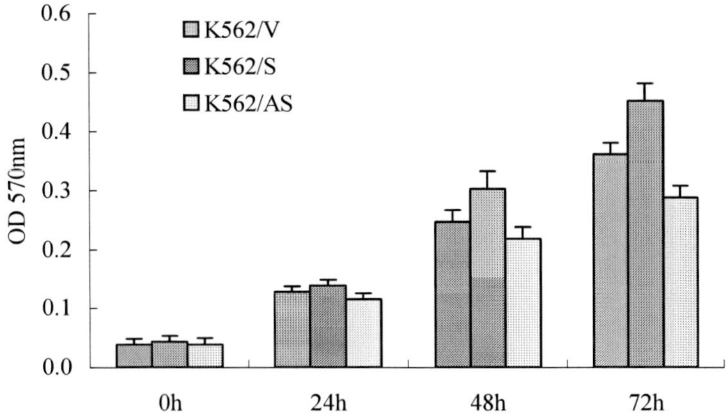

Figure 3. Growth rates of K562/V, K562/S and K562/AS cells by MTT assay (n=20, P <0.01 at 72 hours, K562/AS or K562/S compared to K562/V cells, single factor analysis of variance).

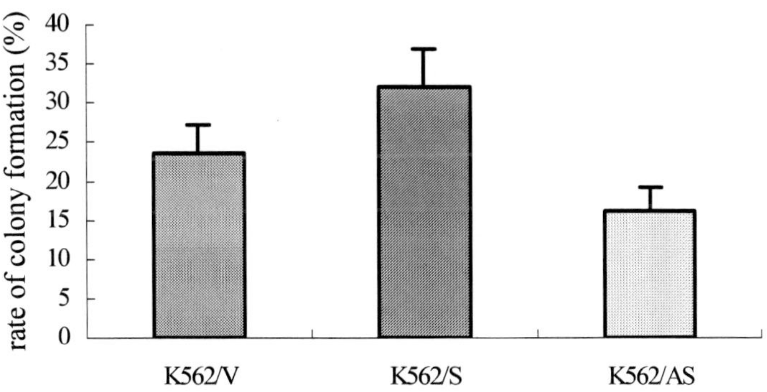

Figure 4. Colony formation efficiency of different transfected K562 cells (n=3, P<0.01, K562/AS cells compared to K562/S cells, single factor analysis of variance).

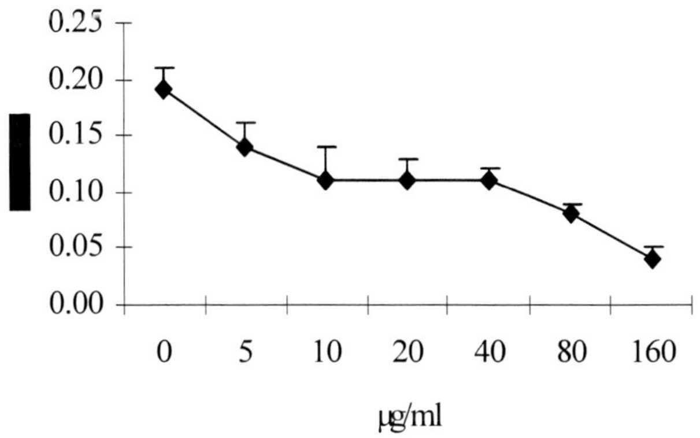

Figure 5. Effects of different concentrations of anti-VEGF monoclonal antibody on growth rate of K562/V cells by MTT assay (n=5).

Table 1. The effect of IFNα and STI571 on the apoptosis of K562/V, K562/S and K562/AS cells (at 72 hours)

Cells		Percentage of apoptotic	cells (%)
	Control	IFNα(100u/ml)	STI571 (0.1μmol/L)
K562/V	11.1±4.7	16.4±2.8	52.8±5.9
K562/S	7.3±3.5	12.4±3.5	42.9±5.5
K562/AS	23.7±2.1*	40.9±2.9**	57.2±6.8

*P<0.05, **P<0.01, compared to K562/V cells (single factor analysis of variance)

The addition of K562/S cell culture supernatants strongly promoted the proliferation of bone marrow endothelial cells (BMECs), more so than the supernatant of K562/V cultures; in contrast, K562/AS supernatant reduced BMEC proliferation as compared to cultures grown with the K562/V supernatant (Figure 6). The bioactivities of K562/S cells could be reduced by pre-neutralization with anti-VEGF monoclonal antibody.

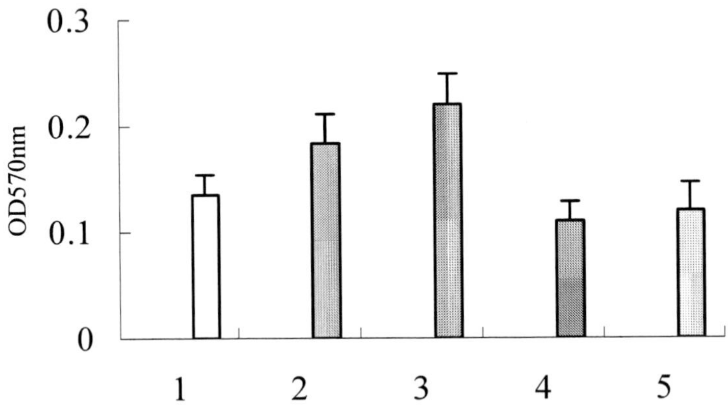

Figure 6. BMEC proliferation rates when cultured for 72 hours in 25% conditioned media from different transfected K562 cell lines. K562/AS supernatant resulted in a marked decrease in BMEC proliferation as compared with K562/V supernatant (p<0.01, single factor analysis of variance). Anti-VEGF monoclonal antibody (0.1 μg/ml) can block the effect of 25% K562/S cells culture supernatant on the proliferation of BMECs (p<0.01, single factor analysis of variance). 1: BMEC medium, 2: 25% K562/V supernatant, 3: 25% K562/S supernatant, 4: 25% K562/S supernatant + 0.1 μg/ml anti-VEGF monoclonal antibody, 5: 25% K562/AS supernatant.

Table 2. Masses of tumors derived from K562/V, K562/S and K562/AS cells in nude mice (mm^3)

Cells	n	+7d	+10d	+14d
K562/V	8	59.3±49.6	131.2±85.1	445.0±150.9
K562/S	6	111.9±62.4	503.7±364.0*	1174.6±508.7**
K562/AS	10	26.8±30.8	72.9±58.40	207.5±192.9*

*P<0.05, **P<0.01, compared to K562/V cells, single factor analysis of variance.

In nude mice *in vivo*, K562/AS cells grew more slowly than K562/V cells, whereas K562/S cells grew more quickly than the control. The tumor masses were 207.5 ± 192.9, 1174.6 ± 508.7 and 445.0 ± 150.9 mm^3, respectively (P<0.05, Table 2, Figure 7). Similarly, the microvessel density (MVD) in masses derived from K562/AS *vs* K562/S *vs* K562/V was 11.0 ± 7.6: 50.8 ± 11.7: 18.9 ± 7.0 (/0.72mm^2), respectively (Figure 8, 9). H & E staining showed that angiogenesis at the borders of the K562/S-derived tumor masses was more than that at the borders of K562/V- and K562/AS-derived tumors.

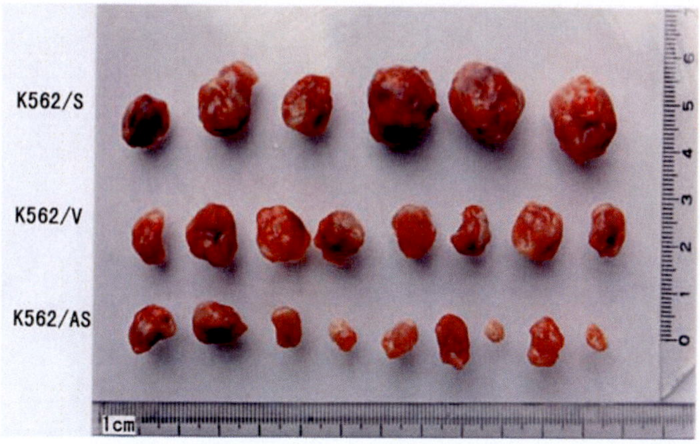

Figure 7. K562/V-, K562/S- and K562/AS-derived tumors grown for 14 days in nude mice.

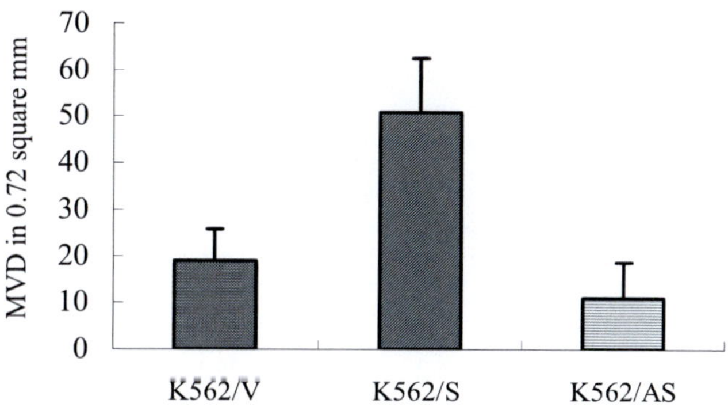

Figure 8. Tumor MVDs from K562/AS cells* are markedly reduced and those from K562/S cells** are markedly increased, compared to those from K562/V cells (*P<0.05, **P<0.01, single factor analysis of variance).

4. Discussion

Vascular endothelial growth factor, a specific mitogen for vascular endothelial cells, has been associated with angiogenesis, growth, metastasis, and poor outcome in solid tumors. Recently, increased bone marrow microvessel density was found in leukemia and lymphoma; reports have correlated this with increased serum levels or high mRNA expression of VEGF [15-22]. We previously reported [13] that the mean VEGF concentration in the plasma from

22 untreated CML patients was 8~9 fold higher than that from 9 CML patients post Allo-BMT (380.6 pg/ml:38.0 pg/ml). In addition, the mean VEGF protein concentration in the supernatant of 17 newly diagnosed CML bone marrow cultures was 2.5-fold higher than that in 11 normal donors (499.8 pg/ml:141.3 pg/ml). These results indicated that CML marrow cells secrete more VEGF than normal marrow cells. However, the effect of VEGF on the abnormal proliferation of CML cells remained unclear prior to this study.

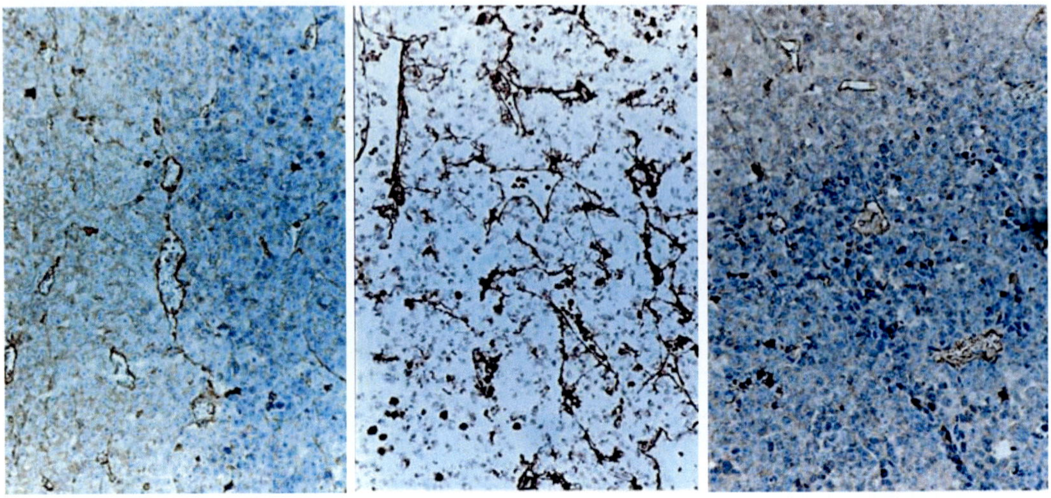

Figure 9. Tumor vascularization (×250) was stained for vWF immunohistochemistry analysis. There is more vascularization in K562/S-derived tumors than in K562/V-derived tumors.

In order to explore the effect of autocrine VEGF on CML cells, K562 cells were transfected with $VEGF_{121}$cDNA sense (S) and antisense (AS) vectors, or the vector (V) alone as control. Our experiments showed that down-regulation of VEGF in K562 cells decreased the growth rate of K562 cells in an autocrine fashion and endothelial cells in a paracrine fashion. These results raised several issues that warrant discussion. First, K562 cells transfected with antisense $VEGF_{121}$cDNA showed reduced VEGF secretion. At present, the only known functions of VEGF are its effect on endothelial-inducing neovascularization and increased vascular permeability; there are no known autocrine effects of VEGF on solid tumors. However, it was recently reported that CML cells express the VEGF-related receptors KDR and Flt-1 [23-25]. This suggests the possibility that VEGF autocrine effects on CML cells may occur through interaction of VEGF and the VEGF-related receptors, KDR and Flt-1. It has been reported that the Flt-1/Fc chimera (a VEGF antagonist) inhibits leukemia and lymphoma cell proliferation *in vitro* [23]. This is consistent with our result that antisense $VEGF_{121}$cDNA-transfected K562 cells showed decreased proliferation, an effect similar to that of cells treated with the anti-VEGF monoclonal antibody. Taken together, these results support the presence of an autocrine pathway between VEGF and its related receptors in hematological malignancies.

A second issue raised by our data is based on our observation that down-regulation of VEGF in K562 cells decreases endothelial cell proliferation. Rimsza *et al.* demonstrated enhanced growth of human umbilical vein endothelial cells (HUVEC) cultured for 72 hours in conditioned media from an acute myeloid leukemia (AML) cell line (HL-60), a Ph+ cell line (K562), and a mantle cell lymphoma cell line (MO-2058), and showed that this effect correlated with the VEGF levels in each cell line [26]. Similarly, our results showed that supernatants from sense $VEGF_{121}$cDNA-transfected K562 cells contained more VEGF and promoted BMEC proliferation, whereas the supernatant from antisense $VEGF_{121}$cDNA-transfected K562 cells did not promote BMEC proliferation (Figure 6). This would suggest that VEGF is perhaps the single most important angiogenesis factor in the studied cell lines. By means of immunohistochemical analysis of tumor specimens from antisense and sense $VEGF_{121}$cDNA-transfected K562 cells, we showed that sense $VEGF_{121}$cDNA-transfected K562 cells have high MVD. As bone marrow angiogenesis is increased in chronic and blastic phase CML [10], it is possible that a VEGF paracrine pathway between CML cells and BMECs may prompt CML cell growth. Possible paracrine loops could include VEGF stimulation of the endothelium to produce stem cell factors, Flt-3 ligand, G-CSF, macrophage colony-stimulating factor (M-CSF), granulocyte-macrophage colony-stimulating factor (GM-CSF), IL-6 and/or IL-7 [27], which, in turn, could act as growth factors for CML cells. Indeed, our results suggest that VEGF is likely to act in the pathogenesis of CML via multiple pathways including both autocrine and paracrine mechanisms.

Several recent reports have indicated that antiangiogenic therapy may be a highly promising new direction in the treatment of hematopoietic malignancies. A neutralizing anti-KDR antibody inhibited 50% of AML cell proliferation *in vitro* and in a SCID mouse model of human leukemia [28], and was associated with prolonged survival of mice xenografted with human leukemia. Similarly, the gene transfer approach used in this report allowed transfected $VEGF_{121}$cDNA to inhibit K562 cell proliferation *in vitro* and *in vivo*. Interestingly, K562/AS was more apoptotic than K562/V and K562/S under same culture conditions. Recent studies have shown that the inhibitor STI-571 inhibited BCR-ABL tyrosine kinase encoded by the Bcr-Abl fusion gene, which is responsible for many of the malignant features observed in CML [29]. STI571 is remarkably effective in treating chronic myeloid leukemia (CML), but the drug is less effective in patients in an advanced phase of the disease called “blast crisis”. In such cases, co-treatment with other chemotherapies enhances the cytotoxic effects of STI 571 against Bcr Abl positive leukemia cells. Here, we found that 0.1 μmol/L STI571 could significantly increase the apoptosis of transfected cells; however, only antisense $VEGF_{121}$cDNA transfection in combination with 100 U/ml IFNα significantly increased apoptosis in K562 cells. This suggests that different apoptotic pathways are initiated by transfection of antisense $VEGF_{121}$ cDNA in the presence of STI571 or IFNα. Specifically, IFNα is an antiangiogenic factor, and there are low serum levels of VEGF in IFNα-treated CML patients [11]. Thus, one of the mechanisms of IFNα action might involve the reduction of VEGF levels. We therefore suggest that the inhibitory effect of VEGF on apoptotic cell death in human hematopoietic cells may involve many proteins, such as ZK7 [30], and that blocking VEGF might enhance the effect of IFNα on CML patients. John *et al.* [31] found that levels of VEGF expression in Bcr-Abl–positive K562 cells were reduced *in vitro* by treatment with STI-571 in a dose-dependent fashion, whereas transfection of Bcr-Abl into human megakaryocyte MO7e hematopoietic cells resulted in enhanced VEGF expression. This further suggests that STI-571 increased apoptosis in K562 cells by inhibiting the Bcr-Abl tyrosine kinase, which further inhibited the levels of VEGF expression. Inhibition of VEGF secretion seemed to partly normalize the abnormal function of K562. Although we cannot directly advocate the development of gene transfer techniques to down-regulate VEGF on the basis of these studies alone, our data suggest that anti-VEGF therapy alone or combination with conventional treatment may be beneficial for CML patients. This could form the basis for new clinical trials and perhaps future innovative treatment for CML.

Acknowledgment

This paper was supported by a grant from the National Nature Science Foundation of China (No. 39970314).

References

[1] Folkman J. Tumor angiogenesis: Therapeutic implications. *N Engl J Med* 1972; 285:1182-6.

[2] Mangi M.H. Angiogenesis and angiogenic mediators in haematological malignancies. *British Journal of Haematology* 2000; 111: 43-51.

[3] Michael K, Patricia AD. Vascular endothelial growth factor and its receptors. *Cytokine & Growth Factor Reviews* 1996; 7 (3): 259-270.

[4] Gerhard S, Georg M-B, Dieter M. The pivotal role of VEGF in tumor angiogenesis: Molecular facts and therapeutic opportunities. *Cancer and Metastasis Reviews* 1998; 17: 241-8.

[5] Folkman J. What is the evidence that tumors are angiogenesis dependent? *J Natl Cancer Inst* 1990; 82(1): 4-6.

[6] Folkman J, Cole P, Zimmerman S. Tumor behavior in isolated perfused organs: in vitro growth and metastasis of biopsy material in rabbit thyroid and canine intestinal segment. *Ann Surg* 1996;164: 491-502.

[7] Srivastava A, Laidler P, Davies RP, Hughes L. The prognostic significance of tumor vascularity in intermediate-thickness (0.76-4.0 mm thick) skin melanoma: a quantitative histologic study. *Am J Pathol* 1988; 133: 419-23.

[8] Weidner N, Semple JP, Welch WR, Folkman J. Tumor angiogenesis and metastasis – correlation in invasive breast carcinoma. *N Engl J Med* 1991; 324:1-8.

[9] Aguayo A, Kantarjian H, Talpaz M, Estey E, Koller C, Estrov Z, O`Brien S, Keating M, Barlogie B, Albitar M. Increased angiogenesis in chronic leukemia and myelodysplastic syndromes. *Blood* 1998; 92 (10, suppl 1): 607a.

[10] Fumiatti R, Soligo D, Venegoni L, Deliliers GL. Bone marrow angiogenesis is increased in CML in chronic and blastic phase, but is completely reversed by bone marrow transplantation. *Blood* 2000; 96 (11): 254b.

[11] Francesco DR, Epifanio F, Maria PA, Raimondo FD, Fichera E, Azzaro MP, Stagno F, Palumbo GA, Giustolisi R. Low serum levels of vascular endothelial growth factor

(VEGF) and hepatocyte growth factor (HGF) in IFN-treated chronic myeloid leukemia (CML) patients. *Blood* 2000; 96(11): 253b.

[12] Aguayo A, Kantarjian H, Manshouri T, Gidel C, Estey E, Thomas D, Koller C, Estrov Z, O`Brien S, Keating M, Freireich E, and Albitar M. Angiogenesis in acute and chronic leukemias and myelodysplastic syndromes. *Blood* 2000; 96(6): 2240-5.

[13] Ruan Guo-Rui, Liu Yan-Rong, Chen Shan-Shan, Li Jin-Lan, Qin Ya-Zhen, Fu Jia-Yu, Bai Ren-Kui. Expression of vascular endothelial growth factor in the bone marrow cells from adult chronic myelogenous leukemia. *Journal of Experimental Hematology* 2001; 9(1): 5-9.

[14] Pruneri G, Bertolini F, Soligo D, Carboni N, Cortelezzi A, Ferrucci DF, Buffa R, Lambertenghi-Deliliers G and Pezzella F. Angiogenesis in myelodysplastic syndromes. *Br J Cancer* 1999; 81:1398-1401.

[15] Perez-Atayde AR, Sallan SE, Tedrow U, Connors S, Allred E, Folkman J. Spectrum of tumor angiogenesis in the bone marrow of children with acute lymphoblastic leukemia. *Am J Pathol* 1997; 150 (3): 815-21.

[16] Veiga JP, Afonso H, Sallan SE, Nadler LM, Cardoso AA. Acute lymphoblastic leukemia cells and their bone marrow stroma collaborate to induce endothelial cell proliferation thereby promoting leukemia cell survival. *Blood* 1998; 92 (10, suppl 1): 511a.

[17] Fiedler W, Graeven U, Ergun S, Verago S, Kilic N, StockschlAder M, Hossfeld DK. Vascular endothelial growth factor, a possible paracrine growth factor in human acute myeloid leukemia. *Blood* 1997; 89 (6): 1870-5.

[18] Padro T, Ruiz S, Bieker R, Burger H, Steins M, Kienast J, Buchner T, Berdal WE, Mesters RM. Increased angiogenesis in the bone marrow of patients with acute myeloid leukemia. *Blood* 2000; 95 (8): 2637-44.

[19] Hussong JW, Rodgers GM, Shami PJ. Evidence of increased angiogenesis on patients with acute myeloid leukemia. *Blood* 2000; 95(1): 309-13.

[20] Shami PJ, Hussong JW, Rodgers GM. Evidence of increased angiogenesis in the bone marrow of patients with acute nonlymphocytic leukemia. *Blood* 1998; 92 (10, suppl 1): 512a.

[21] Chen H, Treweeke AT, West DC, Till KJ, Cawley JC, Zuzel M, Toh CH. In vitro and in vivo production of vascular endothelial growth factor by chronic lymphocytic leukemia cells. *Blood* 2000; Nov 1,96 (9): 3181-7.

[22] Kini AR, Kay NE, Peterson LC. Increased bone marrow angiogenesis in B cell chronic lymphocytic leukemia. *Leukemia* 2000; 14 (8): 1414-8.

[23] Fusetti L, Pruneri G, Gobbi A, Rabascio C, Carboni N, Peccatori F, Martinelli G, Bertolini F. Human Myeloid and Lymphoid Malignancies in the Non-Obese Diabetic/Severe Combined Immunodeficiency Mouse Model: Frequency of Apoptotic Cells in Solid Tumors and Efficiency and Speed of Engraftment Correlate with Vascular Endothelial Growth Factor Production. *Cancer Research* 2000; 60: 2527-34.

[24] Bellamy WT, Richter L, Frutiger Y, Grogan TM. Expression of vascular endothelial growth factor and its receptors in hematopoietic malignancies. *Cancer Research* 1999; 59:728-33.

[25] Lunin S, Kantarjian H, Manshouri T, Verstovsek S, Talpaz M, Faderl S, Cortes J, Albitar M. Clinical relevance of Flk/KDR (VEGF-R2) and VEGF R1 proteins in chronic myelogenous leukemia (CML). *Blood* 2000; 96 (11): 350a.

[26] Rimsza L, Ahrens K, Mainwaring M, Bradshaw K, Trice T, Kao KJ, Braylan R. Both freshly isolated hematologic tumor cells and cell lines stimulate endothelial growth: this effect correlates with VEGF levels only in the cell lines. *Blood* 2000; 96 (11): 60b.

[27] Bertolini F, Mancuso P, Alberto G, Pruneri G. The thin red line: angiogenesis in normal and malignant hematopoiesis. *Experimental Hematology* 2000; 28: 993-1000.

[28] Dias S, Hattori K, Zhu Z, Heissig B, Choy M, Lane W, Wu Y, Chadburn A, Hyjek E, Gill M, Hicklin DJ, Witte L, Moore MAS, Rafii S. Autocrine stimulation of VEGFR-2 activates human leukemic cell growth and migration. *J Clin Invest* 2000; 106(4): 511-21.

[29] Thiesing JT, Ohno-Jones S, Kolibaba KS, Druker BJ. Efficacy of STI571, an Abl tyrosine kinase inhibitor, in conjunction with other antileukemic agents against Bcr-Abl-positive cells. *Blood* 2000; 96(9): 3195-9.

[30] Kuramoto K, Uesaka T, Kimura A, Kobayashi M, Watanabe H, Katoh O. ZK7, a novel zinc finger gene, is induced by vascular endothelial growth factor and inhibits apoptotic death in hematopoietic cells. *Cancer Res* 2000; 60(2): 425-30.

[31] Ebos JML, Tran J, Master Z, Dumont D, Melo JV, Buchdunger E, Kerbel RS. Imatinib Mesylate (STI571) Reduces Bcr-Abl-Mediated Vascular Endothelial Growth Factor Secretion in Chronic Myelogenous Leukemia. *Molecular Cancer Research* 2002; 1:89-95.

INDEX

A

B

C

D

E

F

G

H

I

J

K

L

M

N

O

P

Q

R

S

T